T0038290

*Dr. Joe and What You Didn't Know: 177 Fascinating Questions
and Answers About the Chemistry of Everyday Life*

*That's the Way the Cookie Crumbles: 62 All-New Commentaries
on the Fascinating Chemistry of Everyday Life*

*The Genie in the Bottle: 64 All-New Commentaries on the
Fascinating Chemistry of Everyday Life*

*Radar, Hula Hoops, and Playful Pigs: 67 Digestible Commentaries on
the Fascinating Chemistry of Everyday Life*

SUPERFOODS, SILKWORMS, AND SPANDEX

Science and Pseudoscience in Everyday Life

DR. JOE SCHWARCZ

Published by ECW Press
665 Gerrard Street East
Toronto, Ontario, Canada M4M 1Y2
416-694-3348 / info@ecwpress.com

Cover design: David A. Gee

LIBRARY AND ARCHIVES CANADA CATALOGUING IN
PUBLICATION

Title: Superfoods, silkworms, and spandex :
science and pseudoscience in everyday life / Dr. Joe
Schwarcz.

Names: Schwarcz, Joe, author.

Identifiers: Canadiana (print) 20230572561 |
Canadiana (ebook) 2023057257X

ISBN 978-1-77041-752-6 (softcover)
ISBN 978-1-77852-276-5 (ePub)
ISBN 978-1-77852-277-2 (PDF)

Subjects: LCSH: Science—Popular works. | LCSH:
Pseudoscience—Popular works.

Classification: LCC Q162 .S39 2024 | DDC
500—dc23

This book is funded in part by the Government of Canada. *Ce livre est financé en partie par le gouvernement du Canada.*
We also acknowledge the support of the Government of Ontario through Ontario Creates.

PRINTED AND BOUND IN CANADA

PRINTING: FRIESENS   5 4 3 2 1

CONTENTS

INTRODUCTION

Science, science everywhere . . . it really is. But what is it? The word itself derives from the Latin "scientia," meaning knowledge, which is very appropriate since science can be described as a system of acquiring knowledge through observation and experimentation.

A chemist hoping to produce a biodegradable plastic, a biologist bent on unraveling the mysteries of DNA, or an engineer working on improving a solar panel are all seeking knowledge and are obviously pursuing science. But the scientific quest is not limited to the laboratory. In a sense, we all practice science on a daily basis. Whether we are pondering the relationship between wine and health, questioning why an English cucumber is wrapped in plastic, contemplating the role of inflammation in the body, or just wondering about the possibility of transplanting a human head, what we are actually doing is seeking knowledge. We are pursuing science.

I can't even remember a time when I wasn't interested in science. I have always been intrigued by "what," "why," and "how." What is nylon? Why are antibiotics added to animal feed? How do you make a diamond? I was usually able to find answers even in those pre-Google days, but the search often brought other questions to light. Alright, so nylon is a polymer made by reacting adipic acid with hexamethylene-diamine. That's interesting, but how do we know that? Delving into the subject further reveals that the reaction was discovered in the 1930s by DuPont chemist Wallace Carothers. OK, but how did he think of combining these compounds? And where did he get them? How did he know the product was a polymer? And what is a polymer, anyway? Questions upon questions. What I really came to enjoy was trying to track down the answers.

The path is often a circuitous one, with some fascinating revelation around almost every bend. No matter what the subject, scratching the surface exposes greater complexity beneath. For example, "Lavoisier showed that respiration was equivalent to combustion" may appear as a simple statement in a textbook, but a deeper dive reveals years of ingenious experiments peppered with controversies and human foibles. So it is with almost every discovery. And that is why it is fun to draw back the curtain and look behind the scenes. You never know what you will find. Mostly it is a parade of creativity, cleverness, skill, and originality. Sometimes, though, deceit and fraud rain on the parade. In any case, there are stories to tell. Stories that educate, stories that surprise, stories that entertain, and stories that separate the wheat from the chaff. So let's get going.

BREATHE AND BURN

The guinea pig proved it. Just as Antoine Lavoisier had suspected, respiration and combustion are equivalent processes! Although he is regarded as the father of modern chemistry thanks to his eighteenth-century studies of the role of oxygen in combustion, the discovery of the composition of water, and introduction of a system of naming chemicals, Lavoisier could also sport the mantle of father of the concept of metabolism.

One of the great French chemist's major interests was combustion, a process that had intrigued people since the discovery of fire. He was well aware that a candle burning under a bell jar was eventually extinguished, and he realized this happened because the candle used up a component of air needed for combustion. He named this component "oxygen." The burning candle also released a gas we now know as carbon dioxide, called "fixed air" at the time, that could be absorbed by limewater, a solution of calcium hydroxide. Carbon dioxide reacts with limewater to produce solid calcium carbonate. Scottish physician Joseph Black had shown that "fixed air" was also present in exhaled breath, leading Lavoisier to suspect that respiration was a form of combustion. But how to prove this?

Enter the guinea pig. The French scientist constructed an ingenious device consisting of three chambers with the innermost chamber housing the guinea pig. This was surrounded by a second chamber that held a specific amount of ice, that in turn was surrounded by an outside chamber packed with snow for insulation. The inner chamber was also fitted with a tube containing limewater to absorb any "fixed air" produced. Sure enough, some of the ice melted, the limewater formed a precipitate of calcium carbonate, and the guinea pig began to gasp as the oxygen in the inner chamber began to run out. Indeed, respiration had all the hallmarks of combustion!

Lavoisier did not use the term metabolism, which would not appear in the scientific literature until the mid-nineteenth century and is now understood to refer to the sequence of complex chemical reactions by which the body burns fuel, namely food, to meet its energy requirements. But it was clear to him from the guinea pig experiment that the maintenance of life depended on chemical reactions that involved inhaling oxygen and exhaling carbon dioxide. To explore this further, a human guinea pig was now needed.

A young chemist, Armand Séguin, volunteered. This time Lavoisier designed a mask that was equipped with a carbon dioxide trap and a tube through which oxygen was inhaled. Before donning the tightly fitting mask, the subject was asked to engage in different activities such as eating, exercising, or just sitting in a cold room. The results were clear. After each of these activities, Séguin inhaled more oxygen and produced more carbon dioxide. The implication was that the amount of inhaled oxygen or exhaled carbon dioxide is a measure of the chemical activity, or metabolism, going on in his body.

An obvious conclusion from the experiments with Séguin is that an active person spends more energy than a sedentary one. This principle is described in every book on physiology as well as in the plethora of diet books that flood the market. The message is that exercise will lead to weight loss because the extra energy needed is supplied by the combustion of the body's stores of fat, carbohydrates, or proteins. The loss in weight occurs as the carbon in these body components is converted to carbon dioxide that is then exhaled. Exercise is therefore promoted as a critical partner to diet when it comes to shedding pounds.

Sometimes, however, a beautiful theory can be slain by an ugly fact. Studies by Duke University evolutionary anthropologist Herman Pontzer have cast a huge shadow on the role of exercise in weight control. In the 1950s Nathan Lifson at the University of Minnesota had invented a method of assessing energy expenditure that did not involve wearing cumbersome equipment to monitor inhaled oxygen and exhaled carbon dioxide. Subjects just had to drink "doubly labeled water" in which some of the hydrogen and oxygen atoms are replaced by their isotopes, deuterium and O-18, respectively. Isotopes differ from the common form of the element by having extra neutrons in their nucleus. Incorporating isotopes into a molecule does not alter chemical activity.

After some time urine, sweat, and saliva are collected and tested for the presence of the isotopes using mass spectrometry. It turns out that deuterium ends up exclusively in lost water, but the labeled oxygen turns up both in carbon dioxide and water. Carbon dioxide production, and hence energy expenditure can then be calculated by subtracting the elimination rate of deuterium from that of O-18. Early experiments were burdened by the high cost of the doubly labeled water, but by the early 2000s the price had dropped significantly, allowing Dr. Pontzer to carry out large-scale studies of energy expenditure.

His findings were simply stunning! The hunter-gatherer Hadza people in northern Tanzania, who are extremely active physically, were not expending more energy through their daily routines than sedentary westerners! How could this be? Surely more calories are required for extended romps through the savannah than for watching Netflix on the couch. There can be no doubt about that.

All activities require energy expenditure, including the normal metabolic processes going on even while a body is at rest, known as the basal metabolic rate (BMR). It turns out that when physical activity increases, the body compensates by becoming more biochemically efficient, meaning that it reduces the calories needed to fuel the functioning of the heart, liver, kidneys, and digestive tract. As a result, the Hadza don't "burn" more calories per day than the far less active westerners, they just reduce their BMR so more calories are available for muscular activity. Although the expression "burning calories" is commonly used, it actually makes no sense. A calorie is a unit of measure of energy; it cannot be "burned." Fat, carbohydrates, or proteins can be burned, and they release energy in the process that can be measured in terms of calories. By definition, a food calorie is the amount of heat required to raise the temperature of 1 kilogram of water by 1°Celsius.

Dr. Pontzer's findings also provide an explanation for why dieters at first may lose weight with exercise but then hit a plateau. Like the Hadza, their body readjusts, and the calories needed for exercise are provided by cutting back on calories needed for other body functions rather than by burning fat.

Of course this does not mean that exercise is not worthwhile. The evidence is overwhelming that it is an effective way to reduce the risk of virtually all diseases. But when it comes to weight loss, what matters is what goes into the mouth, not what happens on the treadmill.

BEES AND BANANAS

Swiss entomologist François Huber was blind but that did not prevent him from studying honeybees and publishing his pioneering findings in 1792. He did need some help, of course, and that came from his wife, Marie, and his faithful servant François Burnens who would become his "eyes."

Huber was familiar with the scenario that often plays out after a bee sting and that pain is not the only problem the victim has to contend with. Other bees quickly appear with intent to join the attack. Huber knew that when a bee stings, its stinger gets embedded in the skin and is torn from the bee's body as it struggles to retract it. The outcome is the death of the gallant insect that has sacrificed its life to alert other bees that their home is in need of protection. But how are the support troops attracted? Huber wondered. How do they get the message that there was a potential threat to the colony? Was a clue to be found in the stingers? He instructed his assistant to excise stingers from bees and place them near a hive.

Sure enough, a swarm emerged and headed for the stingers. Did the excised stingers release some sort of odor that was sensed by the other bees? Huber could proceed no further since, at the time, there was no way to determine what specific chemical may be responsible for raising the alarm. The identification of the alarm pheromone would have to wait until 1962, when researchers at Canada's Department of Agriculture noted that the stingers left behind by bees had a sweet scent reminiscent of bananas! That was intriguing enough for a follow-up experiment. Stingers were extracted, macerated, and the resulting solution subjected to analysis by gas chromatography (GC).

GC is an instrumental technique that separates a mixture of gases into individual components as they are being pushed by an inert gas through a column packed with a solid material to which the components bind to different extents. The time it takes for a compound to emerge from the column, known as the retention time, is specific for that compound and is recorded as a peak on a moving chart paper. Each peak represents a different compound, so that a gas chromatograph can determine the number of compounds in a mixture. The invention of this technique is generally credited to a 1952 report by British scientists A.T. James and A.J.P. Martin, although German chemist Erika Cremer had actually described the possibility of constructing such an instrument in a paper submitted to the German journal *Science of Nature* in 1944. The paper was accepted but was not published because the journal's printing press was destroyed during a raid by Allied bombers. Dr. Cremer's article was eventually published in 1976 as a historical item, but by that time credit had been given to James and Martin.

When the Canadian researchers analyzed their chromatogram of the extract of the bee stingers, they found one major peak. Having already noted that the stingers released the scent of bananas, they suspected this peak to be due to isoamyl acetate, a compound that was known to be a major component of banana fragrance. Interestingly, isoamyl acetate had been described as having a banana smell before it had ever been detected in bananas!

By the mid-1800s, chemists had identified various families of molecules and learned to make use of chemical reactions to synthesize novel compounds. For example, when alcohols were reacted with carboxylic acids, they formed "esters" that often had fruity aromas and found application as artificial flavors for candies, beverages, and ice cream. Specifically, reacting isoamyl alcohol with acetic acid produced isoamyl acetate with its potent banana fragrance. This imitation essence delighted the public at the New York City Crystal Palace exhibition in 1853, a hundred years before it was found to occur naturally in bananas!

It wasn't hard to verify that the chromatographic peak in the bee stinger extract was isoamyl acetate. All that was required was the introduction of an authentic sample of this compound into the instrument to determine its retention time. Indeed, it was identical with that of the suspect compound in the bee stinger. Next came confirmation by treating cotton balls with synthetic isoamyl acetate and placing them in front of a beehive. The bees were alerted and became agitated, leaving no doubt that isoamyl acetate was the alarm pheromone of the honeybee!

But how did isoamyl acetate actually trigger aggression in bees? That was answered by a group of French and Australian researchers who managed to untangle the neural mechanism involved. Bees were exposed to an air stream containing isoamyl acetate, were anesthetized, and then quickly frozen in liquid nitrogen to prevent further biochemical activity. They were then dissected and their brain fluids subjected to gas chromatography with the results being compared to bees that had not been subjected to isoamyl acetate. The treated bees had higher levels of the neurotransmitter serotonin in their brain! Could this be the compound that prompts the aggressive behavior?

When a bee's thorax, from where drugs can be readily absorbed, was treated with serotonin, the bee became hostile, and when a serotonin antagonist was similarly applied, it calmed down. The implication is that isoamyl acetate is not the direct trigger that provokes belligerence but rather stimulates the upregulation of serotonin that in turn signals individual bees to attack and attempt to repel a threat. Kudos to the researchers. One suspects that it must take a fair bit of manual dexterity to carry out an anatomical dissection of a bee brain.

Today, synthetic isoamyl acetate is commonly used to impart banana flavor to foods and has to be identified on labels as "artificial flavoring" even though it occurs naturally in bananas. In the future, we may have to rely on it even more given that the Cavendish banana, the one we find in our stores, is threatened by Panama disease, a fungal infection. A wildly popular song, "Yes! We Have No Bananas," released in 1923, may once again become trendy. It was said to have been inspired by Panama disease that at the time was wiping out the Gros Michel banana, destined to be replaced by the Cavendish.

In spite of bananas containing isoamyl acetate, there is no evidence that eating a banana around a beehive increases the risk of being stung. No need to bee-ware.

IT'S ON FIRE!

The May 16, 1868, issue of *The Lancet*, at the time already the premier British medical journal, featured an article with a title that was not quite as stunning as it would be today: "The Holocaust of Ballet-Girls." The term "holocaust" derives from the Greek "holos" for "whole," and "kaustos" for "burnt," explaining why in this case its use was appropriate. The item describes the tragic case of a young dancer who was terribly burned when her muslin dress came in contact with a candle and caught fire during a theater performance. Just one illustration of an occupational hazard peculiar to ballet girls, the article concludes, and goes on to bemoan the negligence of theater managers who ignore the fact that ballet costumes can be rendered fireproof by treatment with sodium tungstate.

The early nineteenth century saw the introduction of gas lighting in homes, city streets, and on theater stages. Footlights on the stage were hailed by audiences, but their open flames presented a risk to performers, particularly dancers who had taken to wearing the novel lightweight but unfortunately highly combustible fabrics being produced. In 1808, English inventor John Heathcote constructed a revolutionary loom capable of weaving fibers into lace that came to be called "bobbinet." This was not welcomed by traditional lace-makers, who worried that the skills they had learned would be replaced by machinery. In 1816, Heathcote's factory was attacked by Luddites, a radical organization of textile workers who opposed the introduction of any mechanical apparatus to produce textiles. (Over time the term "Luddite" came to refer to anyone who opposed the introduction of new technologies.)

Along with bobbinet, muslin, a sheer cotton handwoven fabric, became fashionable in Europe. It was made of delicate yarn and named after the city of Mosul in Iraq where it was first manufactured. The problem was that muslin was flammable! Since combustion requires contact with oxygen, a thinner fabric with more space between fibers is more combustible. A famous 1802 satirical cartoon by James Gillray, undoubtedly inspired by real life cases, depicts a woman reacting in horror as a hot poker from the fire falls on her dress and sets it aflame.

In 1845, at London's Drury Lane Theatre, dancer Clara Webster died when her skirt made of an open-weave net-like fabric called tulle brushed against a gas lamp and caught fire. However, the most famous, and perhaps most tragic, accident occurred on the stage of the Paris Opera in 1862. Emma Livry, the newest star of the opera, was rehearsing a scene when the delicate fabric of her skirt fanned the flame of a gaslight. She suffered extensive burns and died of her injuries eight months later. The real tragedy was that this accident, and many others like it, could have been prevented.

In 1859, an Imperial decree in France required that stage costumes be made fireproof by treatment with a mixture of calcium chloride, calcium acetate, and ammonium chloride in a method developed by French chemist Jean-Adolphe Carteron. Although effective, this "carteronnage" made the fabric stiff and dingy and was disdained by Livry. In a letter to the director of the opera, she agreed to take all responsibility if she were allowed to dance in her untreated costume. The charred remnants of that costume, a grisly reminder of the accident, are now kept in the Paris Opera's museum.

Carteron was not the first to come up with a fireproofing substance. In 1735, Jonathan Wild in England was granted a patent for using alum, ferrous sulfate, and borax. A hundred years later, King Louis XVIII of France commissioned Joseph Louis Gay-Lussac to investigate ways of fireproofing materials. The chemist, best known today for having discovered the relationship between the pressure of a gas and its temperature, found ammonium chloride, ammonium sulfate, and borax to be effective. Then, in 1860, Frederick Versmann and Alphons Oppenheim were granted a British patent for using a number of salts, including the sodium tungstate mentioned in the *Lancet* article, that could have saved many a dancer.

Flammable fabrics were not only risky for dancers. At the end of the nineteenth century, flannelette, a mimic of flannel made of cotton instead of wool, became popular because of its low cost and comfortable feel. It was made by subjecting the surface of the cotton fabric to rollers covered with sharp steel wire to produce minute fluffy fibers that trapped air and created a warm, cozy feel. But the trapped air in contact with the thin fibers set the stage for disaster. Many a child wearing floor-length flannelette sleeping gowns, like those worn by the Darling children in *Peter Pan*, became victims of an epidemic of burns as their apparel was ignited by the flame of a candle or sparks from a fireplace.

The maligned flannelette industry looked for help and consulted William Henry Perkin Jr., son of famed William Henry Perkin Sr., who had founded the synthetic dye industry. The younger Perkin had become a well-known chemist in his own right, and after running hundreds of experiments, found that treating a cotton fabric first with a solution of sodium stannate, then with a solution of ammonium sulfate, resulted in a deposition of stannic oxide within the fabric. This solution made the material fireproof and, critically, was not removed no matter how frequently it was washed with soap and water. By the turn of the century, science had managed to curb the holocaust of injuries and deaths caused by flammable fabrics.

DR. JOE SCHWARCZ

NYLON PROS AND CONS

A tower in the shape of a giant test tube with flashing lights to stimulate bubbling chemicals greeted visitors to DuPont's Wonder World of Chemistry pavilion at the 1939 New York World's Fair. Inside they were treated to a display of nylon, the company's new miracle material "made from coal and air." In 1935, DuPont chemist Wallace Carothers had combined adipic acid and hexamethylenediamine to make "nylon 6,6" with the "6,6" terminology stemming from each component having six carbon atoms. Both are derived from cyclohexanol, which in turn was made from benzene obtained from the distillation of coal tar. The nitrogen needed for the synthesis of hexamethylenediamine came from the air, hence "made from coal and air."

Spectators were amazed as they saw nylon fiber being produced in front of their eyes and then woven into stockings. Next came a tug of war with the stockings to demonstrate the strength of the material. A year after the fair, nylon stockings went on sale to the public with five million being sold the first day! Nylon, touted as the first synthetic material deemed to be better than a natural one, became so popular that hucksters even tried to pawn off silk stockings as nylon.

At the time, there was no concern about the environmental consequences of producing nylon, or about it being made from nonrenewable raw materials. Today, the nylon industry is huge, with "nylon 6" also having entered the picture. That was the brainchild of Paul Schleck at IG Farben in Germany, who, in 1939, based on Carothers's work, polymerized a small molecule containing six carbon atoms, caprolactam, into "nylon 6." Some 18 million tons of combined nylons are now produced every year, with a commercial value of around $10 billion! Toothbrush bristles, lingerie, swimwear, tents, toys, sutures, fishing nets, artificial turf, airbags, reinforcing cord for tires, automobile parts, and "invisible" thread for magic tricks all feature nylon.

The usefulness of nylon is beyond question, but there are questions about the environmental impact of producing the plastic on such an enormous scale. There is the issue of the raw materials being sourced from nonrenewable petroleum. Then there is the problem of nitrous oxide being released when cyclohexanol is treated with nitric acid to produce adipic acid. Nitrous oxide is commonly known as "laughing gas," but its appearance in the atmosphere is no laughing matter. It is a greenhouse gas 300 times more potent than carbon dioxide and accounts for roughly 10 percent of the greenhouse effect. True, nitrogen fertilizer and animal manure are far greater sources of nitrous oxide, but nylon production is a significant contributor with about 30 grams being produced for every kilo of adipic acid.

The adipic acid industry now uses either catalysts or high temperatures to convert nitrous oxide waste to innocuous nitrogen gas, but the ideal would be to produce adipic acid without any formation of nitrous oxide. Enter "synthetic biology." Besides solving the nitrous oxide problem, it can also tackle the issue of using petroleum to make nylon.

Synthetic biology is the manipulation of microbes such as yeasts, fungi, or bacteria with a goal of producing useful chemicals. As a classic example, naturally occurring microbes are used with the production of carbon dioxide by yeast to make dough rise. By the 1970s, scientists had discovered methods to modify DNA (that is, the genetic code of an organism) by inserting a gene from another organism. Bacillus thuringiensis (Bt) is a soil bacterium that produces a toxin fatal to plant-eating insects. The gene that codes for this toxin has been isolated and inserted into the DNA of food crops such as corn and soybeans, thereby enabling them to produce the toxin to deter insects, reducing the need for synthetic pesticides.

This initial form of recombinant DNA technology relied on the use of naturally existing genes. The next goal was the possible synthesis of genes in the laboratory by linking together nucleotides, small molecules that are the building blocks of DNA. By the 1990s, the required methodologies had been worked out and synthetic genes were being inserted into the DNA of bacteria, essentially converting them into little factories to produce the chemicals encoded by the synthetic genes.

Now, getting back to nylon, the target for researchers was a gene that codes for an enzyme allowing glucose to be converted to adipic acid. Inserting this gene into the DNA of a bacterium, *E. coli* for example, would then allow the genetically modified bacterium to convert glucose into adipic acid. Glucose is readily available from plants such as corn, meaning that adipic acid can be produced from a renewable resource, eliminating the need for petroleum. Furthermore, there is no cyclohexanol oxidation involved, so no production of nitrous oxide.

An American company, Genomatica, has now developed a process using synthetic biology to produce not only adipic acid but also caprolactam. It has already manufactured a ton of nylon 6, demonstrating that the technology works. It is now a question of scaling up production, which is underway in a partnership with France's Aquafil. This company's plant in Slovenia has been dedicated to the production of renewably sourced nylon with greatly reduced greenhouse gas emissions.

Nylon production has come a long way since the shortages experienced in World War II when women were asked to give up their nylon stockings so they could be turned into parachutes. Now the challenge is to make those stockings and parachutes in an environmentally friendly fashion.

DÖBEREINER'S LIGHTER AND BERGER'S LAMP

A few years ago, I received a gift after speaking at a cosmetics conference. It was a pretty porcelain item, about the size of a large coffee mug, decorated with artwork and described in an accompanying brochure as a Lampe Berger. From a quick read, I understood that it was to be filled with the fragrant alcohol that was also provided and then lit at the top. I assumed it was akin to a scented candle and stored it away. But now its status has been elevated, and it is on display in my little office museum along with other memorabilia and scientific curiosities.

My interest in the Lampe Berger was rekindled by a comment from Sir Humphry Davy that I came across while researching the history of the Davy safety lamp. I have always been interested in Davy, the first real popularizer of chemistry. In the early years of the nineteenth century, the public flocked to Davy's lectures at London's Royal Institution to be entertained and educated by his chemical demonstrations and accounts of discovery. The remark that got my attention was his description of an observation as "more like magic than anything I have seen; it depends upon a perfectly new principle in combustion." Being interested in magic and in combustion, I had to look into this further.

In 1815, Davy became concerned about the perils that coal miners were facing from explosions of "fire damp," or methane, as we know it today. It is a highly combustible gas released from coal seams during mining, and in days before electric lights the use of candles or kerosene lamps by miners posed a risk of igniting the methane and causing an explosion. Could a lamp be produced that would reduce the risk of igniting the gas? After much experimentation, Davy discovered that a fine mesh made of metal absorbed the heat of a flame and prevented it from passing through. The Davy safety lamp was born with a metal screen surrounding a kerosene-fueled flame.

It was during his experimentation with metals that Davy made his "magical" observation as he experimented with "fire damp" and a platinum wire. Held next to a flame, the wire began to glow as it was heated, which was not a surprise. The surprise came when the flame was extinguished and the platinum wire continued to glow! Somehow, the combustion of methane around the wire was continuing, heating the wire even though no flame was visible. This was the novel type of combustion to which Davy had referred. It was his younger cousin, Edmund Davy, who went on to further investigate this phenomenon. A heated platinum wire exposed to ethanol remained red hot until all the alcohol was consumed! Neither of the Davys realized that they had been experimenting with one of the most important phenomena in chemistry, catalysis. The platinum wire had provided a surface on which combustion occurred without requiring the high temperature of a flame!

In 1821, a German translation of Edmund Davy's account came to the attention of Johann Wolfgang Döbereiner, professor of chemistry and technology at the University of Jena. He repeated the experiment and found that the ethanol was oxidized to acetic acid and noted that the platinum wire was not consumed in the process. Then in 1823 came a pivotal experiment. Döbereiner directed a fine stream of hydrogen gas at a platinum wire a few centimeters away, allowing it to be mixed with air before hitting the target. The platinum quickly became red hot and ignited the hydrogen.

When Swedish chemist Jöns Jacob Berzelius heard about this discovery, he wrote, "From any point of view the most important and, if I may use the expression, the most brilliant discovery of last year is, without doubt, that made by Döbereiner." It was Berzelius who would in 1835 go on to coin the term "catalysis" for the phenomenon, defining it as the "ability of substances to awaken affinities, which are asleep at a particular temperature, by their mere presence and not by their own affinity." Today, we define catalysis as the ability of a substance to increase the rate of a chemical reaction without itself undergoing any permanent chemical change.

Döbereiner managed to capitalize on his discovery. By 1827 he had created the world's first lighter! At the time, if you wanted to light a candle or a kerosene lamp, you needed tinder and a flintstone. Döbereiner designed a device in which hydrogen gas, generated by a reaction between zinc and sulfuric acid, was streamed against a piece of spongy platinum, where it ignited. The flame would then be used to light a candle.

It was the amazing catalytic property of platinum in Döbereiner's lighter that seems to have triggered an idea in the mind of Maurice Berger, a Frenchman with some training in chemistry who was interested in curbing the objectionable smells that plagued mortuaries and hospitals. As early as 1856, there had been a report of the smell of ozone above a heated platinum wire, and there had also been reports of ozone eliminating smells. Berger therefore began to experiment with a flameless alcohol burner, hoping to eliminate odors. He found that his Lampe Berger worked and suggested that ozone was the active agent, although its formation has never been confirmed. It is more likely that smelly compounds in the air come into contact with the hot platinum and are oxidized. Berger lamps are still around and are produced in numerous artistic models and have actually become collectors' items. They are fueled by rubbing alcohol that is combined with a variety of pleasant fragrances.

While there is anecdotal evidence about the lamp's ability to remove smells, proper scientific studies are lacking. There is, however, no question that Lampes Berger make for very attractive collectibles. And for me, they represent a fascinating glimpse into history. I also appreciate that Döbereiner refused to patent his invention, declaring, "I love science more than money, and the knowledge that with it I have been useful makes me happy."

THE DREYFUS BROTHERS' DISCOVERY

Sometimes I'm asked how the ideas for these columns come about. In this case, it was seeing a discarded overhead projector. That brought back memories of my first teaching experience in 1973, when every lecture room was equipped with such a device and we wrote on plastic sheets that we called "acetates." Back then I never gave any thought to that term, but now that unwanted projector sitting in a dumpster somehow made me think of all the chemical reactions I used to write on those acetates. And some of them were all about synthesizing that very substance. "Acetate" refers to cellulose acetate, a material that has played important roles in warfare, photography, fashion, packaging, and medicine.

It all started in 1838 with French chemist Anselme Payen isolating cellulose, the substance that makes up the cell wall of plants. Then in

1865, another French chemist, Paul Schützenberger, reacted cellulose with acetic anhydride to form a gooey substance for which he found no use. But it did interest Camille and Henry Dreyfus, Swiss brothers who had received doctorate degrees from the University of Basel. In 1904, they began to experiment with cellulose acetate in a shed in their father's garden and found that it was soluble in acetone. When a thin layer of this solution was poured on a surface, the acetone evaporated, leaving behind a sheet of a plastic material. An idea was immediately born! Could this be a replacement for celluloid in photographic film that was cursed with the problem of flammability?

Now we need a bit of celluloid history. In 1832, Henri Braconnot combined wood fiber with nitric acid and unknowingly synthesized nitrocellulose, a highly flammable substance he named "xyloidine." German-Swiss chemist Christian Friedrich Schönbein found a more practical formulation for nitrocellulose in 1846 by treating cotton with mixture of nitric and sulfuric acids. He excitedly wrote to Michael Faraday that he was able to shape this material into "all sorts of things and forms" but did not pursue this any further. Schönbein was more interested in his "pyroxyline" burning without producing any smoke, an observation that would lead to the production of smokeless gunpowder.

Then in 1856, Alexander Parkes in England added camphor to nitro-cellulose to produce "Parkesine," the world's first synthetic plastic. Almost simultaneously, John Wesley Hyatt in the U.S. came up with a very similar formulation stimulated by a quest to win a $10,000 prize that had been offered by a billiard ball company to develop a substitute for ivory. It was Hyatt who coined the term celluloid for the novel plastic.

Celluloid had many uses, from replaceable shirt collars and cuffs to combs and flexible photographic film, introduced by George Eastman in the 1880s. But all celluloid products had the same problem. They were highly flammable. Cellulose acetate was not, which is why the Dreyfus brothers thought of using it to replace celluloid in roll film. In 1912, they actually began to produce acetate film but then the fledgling airplane industry got in the way.

The wings on the early biplanes were covered with canvas that when wet created a problem for aviation. A waterproof covering with a lacquer of cellulose acetate, called "dope," solved the problem. The British Cellulose and Chemical Manufacturing Company, along with a similar operation in the U.S., was set up by Camille Dreyfus to manufacture lacquer for aircraft wings. It also proved to be an excellent coating for the fabric used on the zeppelins that were being built in Germany.

After the war, the Dreyfus brothers had another breakthrough. Passing a solution of cellulose acetate through small holes in a showerhead-like device yielded a fiber suitable for weaving into a silky fabric that was widely adopted by the fashion industry. To manufacture the fiber, Camille Dreyfus founded the American Cellulose & Chemical Manufacturing Company in 1918, and then in 1927, purchased the Celluloid Company that had been founded by John Wesley Hyatt and his brother Isaiah. Dreyfus renamed it the Celanese Company of America, with the name "Celanese" deriving from "cel" for cellulose acetate and "ese" because acetate fabrics were so easy to care for. Today, "acetate" still appears on many a label, often blended with silk, cotton, wool, or nylon to produce fabrics that have a smooth feel, are wrinkle resistant, and dry quickly.

The Celanese Company became highly successful, and the Dreyfus brothers became very rich. When Henry died in 1944, his brother created a fund in his memory that was eventually renamed the Camille and Henry Dreyfus Foundation when Camille died in 1956. The foundation administers a number of awards aimed at promoting chemical research as well as chemical education. The biennial Dreyfus Prize recognizes a person who has achieved a major advance in chemistry through exceptional and original research and comes with an award of $250,000. The inaugural prize in 2009 was awarded to Professor George Whitesides of Harvard, one of the most widely cited chemists in the world. He was recognized for his creation of new materials that have significantly advanced the field of chemistry and its societal benefits.

The Dreyfus brothers themselves would be worthy candidates for the prize, were they still alive. Cellulose acetate has unquestionably benefited society in many ways. Just think of all the items that can be traced back to those early experiments by Camille and Henry Dreyfus. Magnetic tape, reverse osmosis water filters, dialysis machine membranes, photographic "safety film," and of course that roll of acetate still forlornly staring back at me from that discarded overhead projector.

SMUGGLING MARGARINE

Live and learn. I now know what the word "caul" means. But let's start at the beginning. This curious journey began with a question I was asked: "How is it possible that they can sell a guacamole dip that has no avocado in the ingredient list?" Never having heard of such a product, I thought I would begin by doing a search for "avocado chemical tricks" to see if perhaps some clever chemist had come up with a way of cutting costs by mimicking the taste of avocado.

That search somehow landed me on a page featuring an 1882 quote by Wisconsin State Senator Joseph V. Quarles. Apparently the senator had thundered: "I want butter that has the natural aroma of life and health. I decline to accept as a substitute caul fat, matured under the chill of death, blended with vegetable oils and flavored by chemical tricks." I now understood why I had landed here, but my eye got stuck on "caul fat." What was that? And what was the senator ranting about?

These days it isn't difficult to hunt down the meaning of a word. It turns out that "caul" is just another term for "omentum," which I know is the fatty membrane that covers the intestines of animals and is sometimes used as a casing for sausages. It seems that it was once also used to make margarine. Now I understood the Senator's ire. He was from Wisconsin, the "dairy state," where cheap butter substitutes were unwelcome.

The imposter was first produced in 1869 by French chemist Hippolyte Mège-Mouriès in response to a financial prize offered by Emperor Louis

SUPERFOODS, SILKWORMS, AND SPANDEX

Napoleon III for the creation of an alternative to butter. The poor couldn't afford butter, and his army had an issue with butter not traveling well. If it wasn't chilled, it would spoil. "An army marches on its stomach," Napoleon I supposedly had declared.

Mège-Mouriès knew that butter was essentially milk fat and began to wonder where the fat came from. Since milk contained fat even when cows were undernourished and were losing weight, he concluded that milk fat came from the cow's body fat. Without enough food, the cows seemed to be sort of draining away. So, the inventive chemist chopped up some beef fat, added milk, minced in some sheep stomach for texture, and cooked the mixture in slightly alkaline water to get "butter." The concoction looked like butter, but it didn't taste great. It didn't have enough "cow" flavor! Mège-Mouriès's remedy was to add some chopped cow udder. That apparently did the trick because in 1870 Napoleon III awarded him the prize and presented him a factory to mass produce the new product. All that was needed now was a name.

In 1813 Michel Chevreul, another French chemist, had isolated an acidic substance from animal fat that formed intriguing pearly drops. He named it margaric acid, from the Greek "margaron," for pearl. Since margaric acid came from animal fat, which was also the source of Mège-Mouriès's discovery, "margarine" seemed a suitable name. It later turned out that margaric acid was not a single substance but rather a mixture of oleic and palmitic acids.

By the 1870s, margarine had arrived in North America, much to the annoyance of the dairy industry. In the U.S., a fierce lobbying effort against the intruder resulted in the passing of the Federal Margarine Act of 1886 that slapped a heavy tax on margarine. The same year, Canada took it one step further and banned the sale of margarine outright. Some states, led by Wisconsin, followed suit and also instituted a ban. Where sales were still allowed, margarine was often colored yellow to make it look more like butter. In 1898, twenty-six states outlawed the addition of color while others took a different approach, requiring margarine to be colored pink to make it more unappealing. This was struck down by the Supreme Court on grounds that enforcing the adulteration of food is illegal!

When the Canadian law was passed prohibiting the production of margarine, Newfoundland was not yet part of the country. Having no large dairy industry, it embraced the production of margarine. In 1925, the curiously named Newfoundland Butter Company, which never produced any butter, was established and began to produce margarine from fish, whale, and seal oil. Being much cheaper than butter, margarine was commonly bootlegged into Canada. When Newfoundland joined the Confederation in 1949, it was with the stipulation that it would be allowed to keep producing margarine. That was granted, although sales to the rest of Canada were prohibited. But just a year later, Canada rescinded the ban on margarine and allowed provinces to regulate sales.

Some provinces required margarine to be bright yellow or orange, while others prohibited any coloring. By the 1980s, most provinces had lifted such restrictions, but Ontario did not allow the sale of butter-colored margarine until 1995. Quebec, the last Canadian province to regulate margarine coloring, repealed its law requiring margarine to be colorless in July 2008.

The laws banning the addition of color made some producers resort to chemical tricks such as the inclusion of a packet of yellow dye with the product. Consumers could then make their own yellow margarine by kneading in the dye, a rather messy business. A more clever idea was the inclusion of a small "color berry" in the plastic bag that contained the margarine. This could be burst without opening the package, resulting in colored margarine with no mess.

Now back to my quest for chemical tricks used to make the "no-avocado guacamole dip." I drew a blank here, being unable to find any such commercial product. But I did find a number of recipes that offered tricks for making guacamole without avocado. Most used young soybeans harvested before they harden, commonly known as edamame, available fresh or frozen. The "trick" is to mash these in a blender with green onions, lime juice, garlic, coriander leaves, olive oil, cumin, and Greek yogurt. Definitely no "caul" is called for. Whether this concoction really tastes like avocado I have no idea, but with the soaring price of avocado may be worth a try.

On the other hand, there is a 2022 study in the *Journal of the American Heart Association* that found a lower risk of heart disease associated with higher avocado intake . . . Hmmm.

THAT'S THE WAY THE RUBBER BALL BOUNCES

Visitors to the Polymer Corporation's pavilion at Expo 67 left with a gift. We walked away with a set of three balls. One ball bounced back to almost the same height from which it was dropped, another was less bouncy, and the third had almost no bounce. They were designed to arouse curiosity, which was the stated theme of the Polymer exhibit. Indeed, inquisitiveness about how things work and how to make them work better is the cornerstone of science. The key to satisfying curiosity is an understanding of the fundamental nature of matter, which in turn is determined by the structure and behavior of the molecules of which it is composed. This is illustrated in an exemplary fashion by the three balls.

Rubber is a polymer, meaning that its fundamental structure consists of long molecules that can be thought of as being built of small molecules, or "monomers," that have been joined together, much like linking paper clips into a chain. Natural rubber, which is an exudate of the Hevea brasiliensis tree, is composed of polyisoprene, a giant molecule formed by linking small molecules of isoprene. These long molecules are coiled in a tangled mess, much like the strands in cooked spaghetti. However, when a stretching force is applied, the molecules are straightened and become more organized. Scientifically speaking they now have less entropy, defined as the state of orderliness of a system. Nature tends to move from organized states to disorganized ones, meaning towards increased entropy. The melting of ice would be a typical example. In ice, the water molecules are in a fixed and ordered state, while in water they are free to move around in a disordered fashion.

When the molecules of rubber are stretched, the entropy of the system has decreased, and when the force is released, the molecules will tend to revert to their disorganized state. The ability of a deformed material to revert to its original shape once the deforming force is removed is known as its elasticity. Natural rubber is not very elastic and has limited use since it gets hard in winter and soft and tacky in summer. Back in 1839, Charles Goodyear was zealously experimenting with improving the properties of rubber and made an accidental discovery that would turn out to be monumental. He had been trying to improve rubber's properties by mixing it with all sorts of substances ranging from soup to nitric acid. On that famously serendipitous day, it was sulfur's turn. Nothing happened, at least not until he accidentally spilled the mixture onto a hot stove. Then the rubber hardened into an elastic mass! Goodyear patented the process, coining the term "vulcanization" after Vulcan, the Roman god of fire. Although he didn't understand this at the time, sulfur atoms had forged links between the long chains of polyisoprene, making them more difficult to untangle and increasing their tendency to return to their original state. Goodyear's exhibit of vulcanized rubber was a major attraction at the Great Exhibition of the Works of Industry of All Nations held in London in 1851. In the impressive Crystal Palace built for the event, Goodyear displayed rubber boats, giant balloons, shoes, medical instruments, and even furniture, all made of vulcanized rubber.

With the introduction of automobiles, vulcanized rubber took on added importance. It was invaluable for belts, gaskets, and, of course, tires. But rubber trees did not grow in Europe or in America, and there was concern about the availability of a product that had to be shipped long distances, especially in case of war. Could rubber be made synthetically? scientists wondered. As early as 1860, English chemist Charles Greville Williams had subjected rubber to destructive distillation and found that it was broken down into isoprene. That began efforts to synthesize rubber from isoprene, a task that proved to be very challenging. However, in 1909, Bayer Company chemist Fritz Hofmann managed to polymerize a closely related compound, methyl isoprene, and produced the world's first synthetic rubber. Then in the 1920s, Sergei Lebedev in Russia polymerized butadiene using sodium to form a synthetic rubber named "Buna" from butadiene and the chemical symbol for sodium, Na. Having learned a lesson from rubber shortages during World War I, Germany embarked on a massive program to produce synthetic rubber, and by the 1930s, chemists at IG Farbenindustrie had developed a series of Buna rubbers, the most famous one being Buna-S, made from butadiene and styrene. During World War II, much of this was manufactured by slave labor at an IG-F factory in Auschwitz.

In Canada, the government established the Polymer Corporation, located in Sarnia, Ontario, to produce synthetic rubber, essentially using the German process, with raw materials obtained from petroleum. After the war, the corporation continued to carry out research in polymerization and it was these efforts that were featured at its Expo 67 pavilion, highlighted by the giveaway of the three balls.

Now, consider what happens when a ball is dropped. The impact with the surface creates a deformation, much like stretching a rubber band. How the ball bounces is determined by how effectively the original shape is restored, which in turn is a function of the specific monomers that have been used to create the rubber and the extent to which the polymer chains are cross-linked, that is, how they are "vulcanized." The ball made of polymerized butadiene restores its shape quickly and bounces high. If the butadiene is polymerized together with styrene, it becomes less bouncy, and with butyl rubber made of isobutene and isoprene there is virtually no bounce. It is all a matter of chemistry, which is the message that was so effectively delivered by the Polymer Corporation's bouncing balls, first raising and then satisfying curiosity. I, for one, got that message. Those balls were instrumental in stimulating my interest in chemistry.

ANTIBIOTIC CONCERNS

They are perhaps the most important class of drugs ever introduced. Antibiotics are the best weapon we have in the fight against disease-causing bacteria. That is a fight we cannot afford to lose. But that may happen if bacteria become resistant to antibiotics.

While antibiotics are very effective at killing bacteria, they are not perfect. Some bacteria are hardier than others, and when a population of bacteria is exposed to an antibiotic, some survive. These will then pass the genetic machinery that allowed them to survive on to their offspring, thereby rendering these resistant to the antibiotic in question as well. Basically, every time an antibiotic is used, there is a chance of developing a strain of bugs resistant to that antibiotic. As a consequence, a subsequent infection caused by that strain will be resistant to antibiotic treatment. The moral here is that antibiotics must be used appropriately, not frivolously, which brings up the topic of their use in animal agriculture.

Cattle, pigs, poultry, and fish can suffer from bacterial infection and require treatment by antibiotics. These are usually administered in the animals' food or water on the advice of a veterinarian who specializes in bacterial diseases of farm animals. No medical intervention is risk free, and with antibiotics we have the dual problems of residues and resistance. A course of antibiotics can result in trace residues of the drug found in the meat after slaughter, and these can then find their way into our body when that meat is consumed. The issue here is not toxicity, but concern that the drug may wipe out susceptible bacteria and allow resistant microbes to thrive. This risk is very small, given that the antibiotic residues in meat are very carefully regulated and treated animals can only be slaughtered after the specific time needed for residues to be eliminated has passed.

A more significant problem than residues is the direct passage of resistant bacteria to people. It is inevitable that an animal treated with antibiotics will develop some resistant bacteria that can then be transferred to humans through contact with animal feces. Farmworkers can become infected and then spread disease. Meat can also become contaminated during slaughter, and traces of fecal matter can be found on meat sold in stores. While heat can kill most bacterial contaminants, inadequate cooking or improper handling before cooking can contaminate surfaces and possibly other foods that are not cooked. Of course, we do not want to consume meat from sick animals, either, so blanket elimination of antibiotics is hardly the answer. Emphasis has to be on the proper and judicious use of these life-saving drugs.

What constitutes an improper use? Feeding antibiotics to animals to make them gain weight more quickly would be one. Although penicillin was discovered by Alexander Fleming in 1928, it was not until the 1940s that it came into widespread use thanks to the work of Drs. Flory and Chain, who managed to isolate the drug from the penicillium mold and developed methods for producing it on a large scale. This also sparked a search by pharmaceutical companies for other antibiotics, and in 1945 researchers at Lederle Laboratories isolated chlortetracycline from a soil sample taken from a field at the University of Missouri. While testing this novel drug, they noted that it caused animals to gain weight. That finding was quickly spun into a commercial enterprise, namely selling antibiotics to farmers who were keen on increased profits by bringing their animals to maturity more quickly. The concept of resistance to antibiotics was not even on the horizon at the time, and the addition of antibiotics to animal feed became a widespread practice.

By the early twenty-first century, the problem of resistant bacteria fostered by the use of antibiotics both in humans and animals was recognized, and in 2006 the European Union banned the use of antibiotics as growth promoters in animals. A decade later both Canada and the U.S. followed suit. With companies no longer able to sell antibiotics to farmers as growth enhancers, they switched to promoting low-dose antibiotics in feed as a way of preventing disease. As a result, there was no significant reduction in the exposure of the animals to antibiotics. Growing concern over bacterial resistance as well as consumer resistance to purchase meat that may contain antibiotic residues has now steered farmers away from the prophylactic use of antibiotics.

Altered practices can also reduce the need for such prophylaxis. For example, piglets naturally wean around three to four months of age, but on factory farms they are often weaned after a month. In this case, they haven't had enough access to antibodies from the mothers' milk, making them more prone to gastrointestinal disease and post-weaning diarrhea. Early weaning also interferes with the development of a healthy microbiome, the proper balance of healthy and harmful bacteria in the animal's gut. A disturbed microbiome can lead to later disease requiring antibiotics. Poultry microbiome is also affected by intense farming practices. Chicks absorb microorganisms through the pores of the egg during brooding, but in modern farming the eggs are taken away from the mother and are cleaned on the surface. Furthermore, once the chicks hatch they do not have the opportunity to go outside and peck away at soil with all sorts of bacteria that would diversify their microbiome and prevent disease. So again, farmers look to prophylactic antibiotics for help. Too bad.

While antibiotic residues in meat are unlikely to have an impact on human health, generating antibiotic-resistant bacteria in animals does pose a threat to people. That's why meat labeled "no antibiotics ever" is enjoying increased sales despite higher cost. However, it must be pointed out that the improper use of antibiotics in humans presents a far greater risk than their use in animals. Pushing doctors to prescribe antibiotics when they are not indicated is tantamount to crying wolf. Should a wolf really come to the door, cries will then bring no help.

SUPERFOODS AND SUPERHYPE

These days you can hardly stroll down the food aisle in a bookstore or browse through a magazine without encountering "superfoods." The term has no legal definition but is usually taken to mean that the food being referred to imparts some sort of health benefit beyond simple nutrition. While the description of a food as being super, from the Latin meaning "above," is relatively recent, the belief that some foods have desirable properties above others is ancient.

As early as 2000 BC, the Chinese deemed garlic to be a digestive aid and the Greeks used it to energize soldiers in battle and enhance performance by the early Olympians. Egyptian pharaohs are said to have provided garlic to builders of the pyramids for extra strength. The famous *Ebers Papyrus*, dating to around 1500 BC, recommends "half an onion and the froth of beer as a delightful remedy against death." Hippocrates recommended lentils as a treatment for ulcers, and the Roman physician Galen in his treatise "On the Powers of Food" described how the body's "four humors" could be affected by diet. The idea that these humors (yellow bile, black bile, blood, and phlegm) were the key to health did not wane until the eighteenth century when James Lind's demonstration of curing scurvy with citrus fruits and Lavoisier's discovery of metabolism laid the foundations to modern nutritional science.

Justus von Liebig's determination of food being basically a combination of fats, carbohydrates, and protein shifted focus from the four humors to the chemical composition of food as a determinant of health. The linking of physiology to diet also saw the emergence of gurus who began to promote specific foods for health. In America, Sylvester Graham advocated a diet of vegetables and coarse grains, and in 1837 even opened a "Graham provision store" in Boston, the country's first health food store. A disciple, James Caleb Jackson, introduced "granula," a bran-rich flour baked and broken into little nuggets that was not only supposed to be healthy but also served to deter people from "self-pleasuring," a practice deemed to be injurious to health. Another Graham devotee, Dr. John Harvey Kellogg, introduced yogurt as a health food, and in the 1940s, J.I. Rodale ascribed wondrous properties to organic agriculture and promoted an array of dietary supplements.

Thus, the foundations for "superfoods" had been laid, but the first use of the exact term is a bit of a mystery. There is a claim that a poem published in a Jamaican newspaper during World War I used the word in reference to wine, and numerous articles about superfoods on the internet refer to a piece supposedly published in Alberta in the *Lethbridge Herald* in 1949 that described a certain muffin as "a super-food that contains all the known vitamins and some that had not been discovered." Maybe so, but a Google search fails to find the Jamaican poem or the muffin article. Bananas took on the mantle of "superfood," without being so-called, with a 1924 article by Dr. Sidney Haas on treating celiac disease in children with a diet of bananas, milk, broth, gelatin, and a little meat. At the time it was not known that the disease was an adverse reaction to gluten and bananas got the credit. The diet worked not because it included bananas but because it excluded gluten.

When it comes to implanting "superfoods" in the public mind, I would argue that the somewhat dubious credit should go to British osteopath and naturopath Michael van Straten, a prolific writer of "natural health" books and host of *Bodytalk*, a long-running radio show. In 1990, he published *Superfoods*, in which he ascribes therapeutic and disease-preventative properties to apples, broccoli, onions, nuts, avocados, and a host of others. He followed up with a number of other "super" books, with the alluring titles *Super Juice*, *Super Soups*, *Superfoods Super Fast*, *Super Boosters*, *Super Herbs*, and for those who don't eat the superfoods, *Super Health Detox*.

Van Straten's ideas about superfoods were germinated by a Swiss health tonic, Bio-Strath, invented by German chemist Dr. Walter Strathmeyer. He began to recommend this to his patients in the 1960s, and based on their reports of enhanced energy and resolution of all sorts of health problems, he started a company to import and market the product. Bio-Strath is a blend of a variety of medicinal plants and brewer's yeast that is rich in the B vitamins. It was this concoction that managed to propel van Straten to fame in a curious way.

Barbara Cartland was at the time already a super famous writer of romance novels, eventually publishing some 723 titles and selling over a billion books. Despite having no scientific education at all, she also ventured into the area of nutrition and described how she took up to a hundred dietary supplements a day. When Margaret Thatcher was prime minister, she received a letter from Dame Cartland with an enclosure of pills that "would take oxygen to every part of the body, including the brain."

In 1964, Cartland wrote an article about being depressed due to the death of her husband. In response, van Straten sent her a couple of bottles of Bio-Strath that spawned a long friendship. The duo even opened an organic health food shop, and when Cartland was asked to be a guest on a radio program about food along with five professors, she only agreed if van Straten could come along. He must have performed well because he was soon offered a regular show of his own that paved the way for his series of super-prefixed books. A deluge of publications by others followed, touting goji berries, noni juice, chia seeds, kale, quinoa, kefir, spirulina, green tea, seaweed, and garlic as being instrumental in keeping the Grim Reaper away.

However, the fact remains that "superfood" is a marketing term, not a scientific one. It is possible to have a healthy diet without including any of the claimed superfoods, and an unhealthy one despite guzzling chaga coffee, maquis berries, or tiger nuts. The only food that can legitimately be called a superfood is whatever Superman eats.

BIOBASED AND BIOBUNK

When it comes to marketing, the prefix "bio" boosts sales. Consumers, increasingly conscious of environmental issues, are attracted by terms like "biodegradable" and "biobased" on product labels. Certainly, the prospect of a conventional plastic bag or discarded water bottle being around for 500 years is not appealing. More comforting is the thought of a biodegradable plastic that decomposes through the action of bacteria or fungi into components that have no adverse impacts on the environment. While there are ways to produce biodegradable plastics, the problem is that they decompose only under ideal conditions and in an unpredictable time frame. Experiments have shown that a supposedly biodegradable bag buried in soil or cast into the ocean undergoes almost no change in three years. "Compostable" plastics are more environmentally friendly, but only as long as they end up in an industrial composting facility.

These days the term "biobased" is appearing on labels with increasing frequency. "Bio" comes from the Greek for "life," so that a "biobased" product suggests the material from which it is made originates from a renewable living source rather than from nonrenewable petroleum. The assumption is that biobased substances have a lower environmental impact in terms of greenhouse gas emissions and have more sustainable supply chains. However, a closer examination suggests that these benefits are quite nuanced.

Surfactants are molecules that have a hydrophilic head that is attracted to water and a hydrophobic tail that favors oily substances. This structure makes them ideal for removing greasy stains from surfaces and accounts for their use in laundry products. Surfactants also find widespread application in cosmetics, where they allow oily and aqueous phases to blend together smoothly. Most surfactants are produced synthetically and source at least some of their components from petroleum. For example, sodium lauryl ether sulfate (SLES), one of the most widely used surfactants, is made from lauryl alcohol and ethylene oxide. The lauryl alcohol is produced from palm kernel oil or coconut oil, while the ethylene oxide is made from ethylene derived from nonrenewable petroleum or natural gas.

Were the ethylene to be produced from plants, then the SLES could be described as "100% biobased." Great for marketing! Indeed some manufacturers of biobased SLES aim to attract their customers, the major detergent and personal care product companies, by offering proof that no petroleum derivatives have been used in their synthesis. That proof lies in the demonstrated absence of any carbon-14 isotope.

When cosmic rays, high energy particles that originate in outer space, bombard Earth's atmosphere, they produce neutrons that can knock a proton out of the nucleus of a nitrogen atom and convert it into the C-14 isotope. The result is that about one in every trillion carbon dioxide molecules in air has a C-14 atom instead of C-12. Since carbon dioxide from the air is the source of all carbon atoms in plants via photosynthesis, living plants will contain some C-14. But carbon-14 is radioactive and has a half-life of about 5,700 years, which means that petroleum formed from living matter at least 65 million years ago no longer contains any C-14. A consequence is that compounds derived from petroleum will not contain any of this isotope, while those originating in living matter will have some.

But "100% biobased" SLES does not necessarily make it more "green." First, although the lauryl alcohol does come from a plant source, there is quite a bit of processing involved. Fats have to be broken down to yield lauric acid, which is then converted to lauryl alcohol by reaction with hydrogen. Both these processes require the use of fossil fuels. That holds true whether the SLES is synthetic or biobased. The other key component, the ethylene needed to make ethylene oxide, instead of petroleum, can be made from ethanol, that in turn is produced by fermenting corn or sugarcane. With both lauryl alcohol and ethylene oxide then being derived from plant sources, the "100% biobased" claim can be sort of justified.

Corn and sugarcane are renewable resources, and petroleum of course is not. However, that does not necessarily mean that biobased ethylene oxide has a smaller environmental footprint. There are significant greenhouse gas emissions associated with growing corn or sugarcane. These crops require pesticides and fertilizer, the production of which relies on fossil fuels, and then there is the fuel needed for the trucks and farming equipment required. Overall, biobased SLES may be associated with somewhat lower greenhouse gas emissions, but the overall impact is not likely to be highly significant unless the ethanol is fermented from waste straw and the CO_2 produced by fermentation is captured.

Instead of a biobased surfactant, how about a "biosurfactant"? Some microbes produce glycolipids, natural molecules that have hydrophilic and hydrophobic parts and can act as surfactants. They can be produced by a specific strain of yeast isolated from honey. While no chemical synthesis is involved, the problem is that the yeast has to be fed raw materials in the form of sugar or sunflower oil. These require the use of agricultural chemicals that have a significant carbon footprint.

What's the bottom line here? That determining the environmental impact of consumer items is complicated and the prefix "bio" does not necessarily mean better. Careful analysis is needed to distinguish between "biobased" and "biobunk."

PICKLED ATHLETES

They still refer to it as "the pickle juice game." It was opening day of the NFL season in 2000 and the Philadelphia Eagles were playing the Dallas Cowboys in Dallas. The day was broiling hot with temperatures on the field hitting 43°Celsius. It wasn't long into the game that some Cowboys players were limping off the field with cramps, but the Eagles were unaffected. In the hottest game in NFL history, Philadelphia triumphed 41–14. Much of the credit went to an Eagles trainer who had suggested the players drink the brine from jars of pickles.

That intrigued Kevin Miller, then an undergraduate student majoring in exercise science at the University of Wisconsin. Studying "pickle juice" was destined to become his passion! It began at Brigham Young University, where the topic for his PhD thesis was "Plasma and EMG responses during an electrically induced muscle cramp and following pickle juice and water ingestion." Dr. Miller, now at Central Michigan University, would go on to become the world's leading expert on pickle juice and carried out the most frequently cited study on the subject, one that actually indicated the Eagles may have been on to something. Perhaps not Nobel Prize material, but the results were welcomed by athletes for whom cramping is a curse. After all, it is not unusual to see a basketball player writhing on the floor with a cramp.

Pickle juice is really a misnomer. Unlike oranges or apples, pickles are not squeezed to produce juice. The reference is to the brine in which cucumbers are fermented for conversion into pickles. That conversion is quite simple and has been known for thousands of years. Just submerge the cukes in salty water and wait three to four weeks. Voilà! Garlic, dill, and mustard seed can be added for flavor, but these are not necessary for pickling, which is probably the oldest method of food preservation.

Necessary, however, are salt and bacteria that naturally inhabit the surface of the cucumber, having been picked up from the air or the soil. The most important are from the family of lactobacilli because these produce lactic acid, the key for preservation. There are many other types of bacteria that colonize the cucumber, some of which can lead to spoilage or even illness. Fortunately, these are inhibited by salt to a far greater extent than the lactobacilli. When it comes to the battle of the microbes, as long as the solution in which the cucumbers are immersed is sufficiently salty, the lactobacilli win. They multiply quickly and digest the carbohydrates in the cucumber to produce lactic acid which increases the acidity of the solution, producing the desired tart flavor. More importantly, the increased acidity prevents other less desirable bacteria from multiplying.

Lactobacilli need an oxygen-free, or anaerobic, environment to grow, while other bacteria can multiply in the presence of oxygen. That is why it is important to exclude air while the fermentation is going on. Any exposed pickle or brine becomes a breeding ground for microbes that will spoil the whole batch.

So, what's in pickle juice? Lots of salt. Also some lactic acid that leaches out from the fermented cucumbers along with small amounts of potassium, magnesium, and calcium. Then of course there are also the lactic acid bacteria, which are in the realm of "probiotics," defined as microbes that have beneficial effects when introduced into the body. But can this concoction really help resolve a cramp?

The common belief used to be that cramps are caused by a combination of dehydration and loss of sodium and potassium. This has stimulated athletes to guzzle sports drinks like Gatorade, but a clever experiment by Dr. Miller showed that the cause of cramps is more complicated. He devised a way to trigger cramps in the big toe through electrical stimulation and had volunteers pedal on a semi-recumbent exercise bike to a point of dehydration. But it took no less electrical stimulation to produce a cramp than before they had exercised. Dehydration didn't prime them for a cramp.

When cramps were induced in volunteers after they had become exhausted from bicycling, they lasted about two and a half minutes. They were then zapped again, and as soon as the cramps began, the men drank 75 milliliters either of deionized water or pickle juice from a jar of Vlasic pickles. This time cramps lasted only about 85 seconds in the subjects who drank the pickle juice, leading to the widely reported result that "pickle juice relieves a cramp 45 percent faster than drinking no fluids, and 37 percent faster than water."

The relief was so rapid that the juice had hardly enough time to reach the stomach. Consequently, the effect could not be explained by the restoration of electrolytes that are found in pickle juice, with sodium being prevalent. Rather, the researchers suggested that the pickle juice may trigger a reflex in the mouth that sends a signal to inhibit the firing of motor neurons in the cramping muscle. As far as pickle juice actually improving performance, as some tennis players claim, nope, says Dr. Miller, based on an experiment in which young men drank either deionized water or pickle juice before running to a point of exhaustion. There was no difference between water and the "juice."

Some consumers have raised questions about the safety of drinking pickle juice, or even just eating pickles, given that the International Agency for Research on Cancer (IARC) ranks pickled vegetables as "possibly carcinogenic to humans." Some fungi in these vegetables can turn naturally occurring nitrates into nitrites that then form carcinogenic nitrosamines. But fret not, this refers to Asian diets in which pickled vegetables may be eaten every day as a staple food. That crunchy dill you have along with your occasional smoked meat sandwich isn't going to cause a problem. Be mindful, though, of the sodium content. If you have a protracted high-sodium diet, that, one might say, can land you in a pickle.

FASCINATING FIBERGLASS

More than 27 million people flocked to Chicago's Columbian Exposition in 1893, a six-month celebration that honored the arrival of Christopher Columbus to the New World. They came to see replicas of Columbus's ships, to ride the original Ferris wheel, and to watch Harry Houdini and his brother Theo perform at the fair's midway. They also came to ogle at the Libbey Glass Company's famous glass dress! If they thought the dress would be transparent, they were disappointed. It wasn't. The dress was opaque, actually woven of silk and thin fibers of glass. It was destined to be no more than a curiosity since it was heavy, and due to the easy breaking of the glass fibers, rather prickly. But just seventy-six years later, fabric made of glass fibers would hit the news again as astronauts walked on the moon protected by space suits with an outer layer of glass fibers.

Glass fiber, as the term implies, refers to a thin thread made of glass. I was familiarized with such fibers back in my graduate school days when sometimes there was a need to bend glass tubing into a desired shape. This can be done by softening the glass in the flame of a Bunsen burner, but if instead of bending, the heated glass is quickly pulled, it forms a thread. We used to play around to see who could produce the longest and thinnest thread. With a little practice, you can produce a fiber that is thin enough to be woven into fabric. That fabric would be referred to as "fiberglass."

Today, of course, this is all done by machines, and threads can be produced that are thinner than a human hair. When matted together these fibers can trap air, making for an excellent insulating material that is commonly used in walls and attics. In this case, description of the material as "fiberglass" is accurate because it is made of nothing but glass. However, confusion can arise since "fiberglass" is also commonly used to describe a composite material made by impregnating a network of glass fibers with a fluid resin that then "cures" to form a hard substance.

This technology was first worked out by German chemists in the late 1930s. They discovered that polyester resin can be cured by combining it with styrene and a hydrogen peroxide "initiator." The peroxide triggers a reaction that allows the styrene to crosslink the long polyester molecules to form a rigid network. While the mixture is still fluid, it can be poured into molds where it will then harden as the crosslinking reaction proceeds.

During World War II, British intelligence agents were successful in stealing the secret for this reaction from the Germans and turned it over to Cyanamid, an American company. It wasn't long before airplane parts, panels for ships, and domes to protect radar equipment were being manufactured. After the war, "glass reinforced plastic" found its way into fishing poles, pleasure boats, and in 1953, into the body of Chevrolet's Corvette.

When Alan Shepard, America's first astronaut, was launched into space in 1961, he was sitting in a fiberglass seat custom-molded to his body. In his *Mercury* capsule he was protected from the heat of re-entry by a heat shield consisting of an aluminum honeycomb covered with multiple layers of fiberglass. The *Apollo* capsule that would take astronauts to the moon, as well as the lunar lander, were insulated with fiberglass.

The journey to the moon required extensive testing of the *Apollo* capsule on the ground before the first low Earth orbital test, planned for 1967. Tragically, that launch never took place because astronauts Gus Grissom, Ed White, and Roger Chaffee were killed in a fire that engulfed the capsule during a ground test. The atmosphere in the capsule had been designed to be 100 percent oxygen to save weight, and while oxygen does not burn, it supports combustion. An electrical spark triggered the flash fire that was fed by combustible materials such as Velcro, extensively used in the capsule.

The astronauts had been wearing fireproof suits made of DuPont's Nomex, but it could not stand up to the intensity of the flames. NASA launched a full-scale investigation and tasked companies to come up with a superior material. That challenge was met by the Owens Corning Company with "Beta cloth," made of tightly woven, extremely thin glass fibers coated with Teflon. This was totally non-flammable, had a higher melting point than Nomex, and the tight weave prevented penetration by gases or microscopic particles. It was ideal for the outer layer of the *Apollo* space suits.

Walter Bird, an engineer who in the 1940s had worked on designing coverings for radar installations, saw the potential of "Beta cloth" in Earthly applications. In 1975, his company Birdair installed a roof made of panels of Teflon-coated fiberglass on Detroit's Pontiac Silverdome. This then forayed into similar installations around the world, including the sail-like structures that top Vancouver's Canada Place, the covering of the Dallas Cowboys Stadium, and the roof of Montreal's Olympic Stadium.

The original roof for the stadium, built for the 1976 Olympics, was designed to be retractable and was made of Kevlar, DuPont's famous bulletproof material. Unfortunately, it did not stand up well to the rigors of opening and closing and was replaced in 1998 by a non-retractable roof constructed of panels of Birdair's fiberglass. Although the fabric stood up to the demands of outer space, it could not deal with Montreal's snowfall. The weight of snow that piled up on the roof caused numerous rips, and the city is once again looking for a contractor to design a new roof. Given that chemists and engineers were able to solve the monumental problems involved in putting men on the moon, one would think they should be able to put a roof on our Olympic Stadium. Then we just have to find a team to play under it.

"SOOTHING, QUIETING, AND DELIGHTFUL BEYOND MEASURE"

That was Queen Victoria's description of chloroform, the anesthetic administered by her physician Dr. John Snow to ease labor pains as she gave birth to Prince Leopold in 1853. Dr. Snow was Britain's leading anesthetist at the time, having followed in the footsteps of William Morton, who in 1846 had introduced ether anesthesia in Boston. Within a year, Snow had designed an ether inhaler and published *On the Inhalation of Ether*, a practical guide for the administration of the drug.

That same year, Scottish physician James Simpson discovered the sleep-inducing properties of chloroform. After dinner every night, Dr. Simpson and two assistants had the habit of experimenting with various chemicals to see if they had any anesthetic effect. Chloroform was a winner! Within days, Simpson had used it to perform minor surgeries, and when his patient Jane Carstairs experienced severe labor pains, he seized the opportunity to try chloroform for childbirth. A handkerchief soaked with chloroform placed over Jane's mouth induced sleep and the unconscious mother proceeded to deliver a healthy baby! Some accounts claim she was so thankful that she named the baby Anaesthesia. Would be a great footnote to the story if it were true. It isn't. The baby was named Wilhelmina.

Upon hearing of Simpson's success, Dr. Snow investigated chloroform as he had done for ether and found the drug to be more potent, but also potentially more dangerous. In 1848 he learned of the death of fifteen-year-old Hannah Greener, who had been anesthetized with chloroform in preparation for minor surgery for an ingrown toenail. Within minutes Hannah became pulseless and died. Snow concluded that chloroform had to be administered in a carefully controlled fashion and engineered a vaporizer capable of delivering measured amounts of the drug. Queen Victoria, who had struggled through seven previous births, and had referred to pregnancy as "the shadow side of marriage," reaped the benefits.

The administration of an anesthetic for childbirth was not without controversy. Some physicians believed the relief of pain would slow the progress of labor, and the Church of England opposed easing pain in childbirth on theological grounds. Eve's original sin doomed all women to bring children forth in pain! Opposition, however, was subdued after Victoria gave birth to Leopold, and later to Princess Beatrice, under chloroform anesthesia. If "chloroform à la reine" was good enough for Her Majesty, then surely it was good enough for all.

Most accounts of the history of chloroform begin with James Simpson's self-experimentation and do not raise the question of where the chloroform came from. Delving into this makes the story even more interesting.

At the end of the eighteenth century, a group of wealthy Dutch amateur scientists formed the Batavian Society to study the new science of chemistry. In 1794, they reported treating alcohol with sulfuric acid to produce a gas that when bubbled into a chlorine solution yielded an oily fluid that came to be called "Dutch oil." Today we know that the reaction of alcohol with an acid yields ethylene, and that ethylene reacts with chlorine to form dichloroethane. "Dutch oil" was dichloroethane, but of course, this was not known at the time.

Alcohol was available to the Batavians from fermentation, and sulfuric acid had been known since about the eighth century when alchemist Jabir ibn Hayyan heated "green vitriol" (iron sulfate) to produce "oil of vitriol," the original term for the acid. And chlorine? That was discovered in 1774 by Swedish chemist Carl Wilhelm Scheele, who treated hydrochloric acid with the mineral pyrolusite (manganese dioxide).

In 1820, Glasgow physician Thomas Thomson found that an alcoholic solution of "Dutch oil," for which he coined the term "chloric ether," was a stimulant. American physician and amateur chemist Samuel Guthrie heard about this, and in 1831 thought he had found a simpler and cheaper way to produce the substance by reacting whiskey with chlorinated lime (calcium hypochlorite). His sole evidence was that his product smelled like chloric ether and when inhaled produced a pleasant stimulating effect. He was wrong. What he had actually produced was chloroform.

Simpson's legendary self-experimentation began when he learned about Morton's introduction of ether as an anesthetic. Could he find something that worked even better? He tried chloric ether but complained to his former medical school friend David Waldie that it didn't work well. Waldie, who had given up medical practice in favor of chemistry, was familiar with chloric ether because he had distilled chloroform from it. Now he suggested that Simpson try pure chloroform.

The question was where to get the chemical. Simpson approached Edinburgh chemists William Flockhart and John Duncan who had a small pharmaceutical firm. The two worked overnight to produce a pure sample of chloroform, and the rest, as they say, is history. Duncan Flockhart & Co. became the prime manufacturers of chloroform and supplied the British and Allied forces during the two World Wars. It was the company's chloroform that John Snow administered to Queen Victoria.

Chloroform was not problem free. On occasion, it was found to cause respiratory failure and cardiac arrhythmia. When better anesthetics were developed, chloroform faded into the background. But it was still used in 1953 when I had my tonsils removed. I vividly recall the doctor pouring some on a gauze pad that was then clamped over my mouth. I was told that my tonsils were plucked out in a minute and a half, apparently setting a new record at the time. I would next encounter chloroform when I spilled some on my hand in the lab. The burning sensation between my fingers was not a pleasant experience.

Although Dr. Snow made highly significant contributions to the practice of anesthesia, he is actually better known for his discovery that cholera can be transmitted by contaminated water. His tracing of a cholera epidemic to water drawn from a pump in London's Broad Street and his advice to remove the pump's handle to stop the epidemic are regarded as the world's first successful flirtations with epidemiology.

FROM "SWILL MILK" TO PASTEURIZATION

Isabella Beeton certainly did not intend to harm children. But a casual remark in her wildly popular 1861 book, *Mrs. Beeton's Book of Household Management*, about boracic acid purifying milk was responsible for many children getting sick and even dying from drinking milk contaminated with bovine tuberculosis bacteria.

With the advent of the Industrial Revolution, and the growth of cities in Victorian England, urban dairies could not keep up with the demand for milk. Unfortunately, transportation from farms in those pre-refrigeration, pre-pasteurization days allowed time for various microbes in the milk to multiply. Some of these microbes produce enzymes that convert the lactose and proteins in milk into smelly, foul-tasting compounds, and farmers had somehow discovered a work-around. Adding boracic acid, a mixture of sodium borate (borax) and its acidified derivative, boric acid, countered the milk's unpleasant smell and off-taste. Mrs. Beeton was aware of this and assured her readers that boracic acid was harmless, and recommended that they could even preserve their milk longer by adding some themselves. Bad idea!

Boron compounds are not harmless. Indeed, an article in 1887 in *The Lancet*, the prime medical journal at the time, claimed that "even small quantities of boracic acid are capable of exerting a distinctly injurious action on the human organism." This, however, was not the major problem. While boracic acid retarded the growth of microbes that caused the unpleasant sensory properties, it did not prevent the growth of disease-causing organisms. The longer the milk was kept, and boracic acid allowed for that, the more time the tuberculosis bacteria had to multiply. Infection of Victorian children with bovine tuberculosis was common, and many deaths could have been spared had boracic acid not contributed to the illusion of safety.

Milk was killing children on this side of the pond as well. Here, the problem wasn't the masking of spoiled milk with boracic acid, it was contaminated milk from cows kept in filthy conditions and fed brewery waste "swill." New York, like London, had become a huge bustling city with an ever-increasing demand for milk to feed infants. Distilleries in the city produced large amounts of alcoholic mash left over from making whiskey, something that did not go unnoticed by milk producers. Dairies sprang up around distilleries where cows were fed the swill. To maximize profits, the animals were squeezed into narrow stalls where they became covered with flies and wallowed in their own excrement. No wonder they became so sick they could hardly stand. Even the milk they produced looked sickly. To alter its bluish color, plaster of paris and molasses were added, and flour was used as a thickener. A *New York Times* editorial in 1858 described "swill milk" as a "bluish-white compound of true milk, pus and dirty water" produced by "running distillery slops through the udders of dying cows and over the unwashed hands of milkers." The *Times* estimated that every year some 8,000 infants died from drinking swill milk.

Now, let's hop back to Europe, where in 1856 Louis Pasteur, then professor of chemistry at the University of Lille, was investigating why wine sometimes turns sour. Looking through a microscope, he noted the presence of bacteria that he then determined were capable of converting alcohol into acetic acid. These bacteria, Pasteur found, could be inactivated by heat. This inactivation became known as "pasteurization." Pasteur was not the first to note that beverages or foods could be preserved with heat. Some forty years earlier, Nicolas Appert had shown that boiling food in a glass jar that was then sealed prevented the contents from spoiling. Pasteur's contribution was working out conditions that allowed for heating at a lower temperature for a shorter time to preserve texture and flavor. He did not work with milk. It was Franz von Soxhlet, a German agricultural chemist, who first suggested in 1886 that milk sold to the public be "pasteurized."

Back across the ocean once more. Nathan Straus, who as a young boy had emigrated from Germany to America, grew up to become the wealthy co-owner of Macy's department store. He became interested in milk when he learned that a cow on a farm he owned had died of tuberculosis despite appearing to be healthy. Could milk from such cows be entering the marketplace and making children sick?

Upon learning von Soxhlet's suggestion to "pasteurize" milk, Straus became a passionate advocate. In 1893 he built the Nathan Straus Pasteurized Milk Laboratory and set up stations in poor areas of New York to give away milk, eventually establishing 297 milk stations in thirty-six cities, all at his own expense. Nathan Straus is estimated to have directly saved the lives of close to half a million children!

Straus was a great philanthropist, writing in his will that "what you give for the cause of charity in health is gold, what you give in sickness is silver, and what you give in death is lead." He had a long, healthy life, gave away millions, built shelters for the homeless, distributed food and coal to the poor, and founded a health center in Jerusalem, which he said was to be for all the inhabitants of the country, irrespective of race, creed, or color. The Israeli city of Natanya is named after him.

The pasteurization of milk has been one of the most successful public health measures of all time, but has not escaped criticism. Advocates of drinking "raw milk" claim that pasteurization destroys vital nutrients in milk and that there is no problem with drinking raw milk from healthy cows. Farmers who produce raw milk, they say, are better at taking care of their cows than factory farmers.

I would rather play it safe and stick to pasteurized milk. As for borax? It belongs in the washing machine, not in milk.

FRYING WITH WATER

Telling students that you are going to fry an egg in the lecture room doesn't trigger much excitement until you tell them that you are going to do it with water. You then proceed to pour some water into an aluminum pie pan and cover it with another empty pie pan. An egg broken into the top pan will start to fry within a few minutes. Magic? In a way, yes. The magic of chemistry.

The "trick" is placing some calcium oxide, or quicklime, in the bottom pan. While we recognize water as an extremely important resource when it comes to cleaning, cooking, irrigation, or serving as a solvent for chemical reactions, we don't generally think of it as a reagent in chemical reactions. Yet that simple H_2O molecule can engage in a number of very useful chemical reactions. Indeed, without one of these reactions we would not exist. Photosynthesis, upon which all life depends, is the reaction of water with carbon dioxide to produce glucose and oxygen.

There's more. A key step in the production of sulfuric acid, the world's most important industrial chemical, is the reaction of sulfur trioxide with water. Hydrogen gas, vital for fertilizer production, is made by reacting methane with water.

Calcium oxide has an extreme thirst for water and reacts with it very quickly, hence the term "quicklime." The product of the reaction is calcium hydroxide, or "slaked lime," so called because quicklime's thirst for water has been "slaked." This is a highly exothermic reaction, producing enough heat to quickly fry an egg. It is also the technology used in self-heating cans of food or drink. These cans have two chambers with one housing whatever is to be heated while the other contains calcium oxide separated from water by a barrier. Pushing a button on the can breaks the barrier and heat is quickly generated as the quicklime reacts with the water.

The ability to generate heat is far from the main use of calcium oxide. It is critical for the production of cement, which when combined with gravel and sand yields concrete, the most widely used material in the world. Calcium oxide is produced by heating limestone (calcium carbonate) and is then fired with clay to make cement. Since heating limestone releases carbon dioxide, and the high temperatures needed to react calcium oxide with clay require burning fossil fuels, the production of cement has a colossal carbon footprint. Roughly 8 percent of all global carbon emissions caused by humans is due to cement. But life without cement is unimaginable.

The setting of cement involves a series of complex chemical reactions that are only partly understood, but the key is the reaction of quicklime with water to form slaked lime that then slowly absorbs carbon dioxide from the air to form insoluble calcium carbonate. Although this does remove some carbon from the air, it does not compensate for the amount of carbon dioxide released in the making of cement.

While it is clear that water can act as a reagent in some extremely useful reactions, it can also initiate some reactions that can have deadly consequences. The world's worst industrial accident, the Bhopal tragedy, is a case in point. Many aspects of the disaster are debated to this day, but there is no doubt that on the evening of December 2, 1984, a massive amount of highly toxic methyl isocyanate gas (MIC) was released from a Union Carbide plant in the Indian city of Bhopal. It quickly enveloped the surroundings and resulted in the immediate death of over 3,000 people and the eventual demise of more than 15,000. Close to half a million others suffered injuries ranging from blindness to chronic bronchitis.

Liquid methyl isocyanate was stored in a large tank connected to a pipe through a valve that when opened allowed it to be released into a chamber where it would react with naphthol to produce the pesticide carbaryl. Accidental introduction of water into the MIC tank seems to be the most likely explanation for the disaster. Apparently, one day an untrained technician failed to properly shut the valve as he was cleaning the pipes with water. Methyl isocyanate reacts quickly with water to form dimethylurea and carbon dioxide. While these are not toxic, the reaction is extremely exothermic and caused the MIC in the tank to boil. The increased pressure then burst a safety valve, resulting in some 40 tons of toxic MIC vapor being spread over most of Bhopal.

The plant had been designed with a number of safety systems, but these either failed or had been made inoperative due to costs. A refrigeration system to cool the MIC tank and a sodium hydroxide scrubber had both been shut down. The latter was designed to neutralize any escaping methyl isocyanate by reacting with it to produce innocuous products, but it had been turned off as a cost-cutting measure. Union Carbide claimed, without any real corroborating evidence, that the tragedy was not an accident but an act of sabotage.

After much legal wrangling, the company agreed to pay $470 million to set up a fund for the victims of the disaster and to build a hospital in Bhopal dedicated to their treatment. The terrible tragedy resulted in the institution of stringent safety measures for the production of chemicals of all kinds, but without doubt has left a huge stain on the chemical industry. A sad part of the story is that carbaryl can be produced by a method that does not require methyl isocyanate at all. However, it is more expensive, and the company chose the cheaper route.

A final note. When methyl isocyanate reacts with water, the products that form are not harmful. Had the population in the area been told to just cover their head with a wet towel in the event of a chemical leak from the plant, many injuries, particularly to the eyes, would have been avoided. Water obviously can be a useful or a sometimes dangerous reagent.

A LEGENDARY NEON SIGN

Ask a Montrealer to name the city's iconic landmarks and you will likely hear about St. Joseph's Oratory, Schwartz's Deli, the Olympic Stadium, and the "Five Roses" neon sign. That sign has been part of the city's skyline since the 1940s and actually used to read "Farine Five

Roses Flour" until 1993 when the word "Flour" was removed, believe it or not, because it was English. But that sign with its fifteen-foot-high letters still glows bright red every night and prompts a journey through the history of neon.

That journey begins in 1785 with "natural philosopher" Henry Cavendish, as scientists were called at the time, discovering that a tiny amount of gas remained when "phlogisticated air" (nitrogen) and "dephlogisticated air" (oxygen) were removed from a sample of atmospheric air. Cavendish was unable to identify the gas and it remained a mystery for another one hundred years until William Ramsay and Lord Rayleigh became interested in the problem. They passed air over red-hot copper to remove oxygen as copper oxide, and then over hot magnesium to remove nitrogen as magnesium nitride. A tiny amount of gas, roughly 1 percent of the original, still remained, and it had a curious property. It would not engage in any chemical reactions! They named this residue "argon" from the Greek for "inactive" or "lazy."

Ramsay, with colleague Morris Travers, managed to liquify this residue by cooling it, and discovered that when it was slowly heated, tiny amounts of other gases boiled off. These turned out to be a series of new elements that they named neon, krypton, and xenon from the Greek for "the new one," "the hidden one," and "the stranger." They came to be known as the "noble gases" because, like nobility, they had no tendency to associate with commoners.

Some forty years before Ramsay identified the noble gases, German glassblower Heinrich Geissler had managed to use a vacuum pump to partially evacuate a glass tube. When he applied a high voltage to the electrodes that had been fitted to the ends of the tube, the inside of the tube began to glow. At the time there was no explanation for this phenomenon, which is actually a consequence of trace amounts of gases present in the tube. A rationale is to be found in modern quantum theory that describes how electrons in an atom can exist in different energy states. When they absorb electrical energy they become excited, and on returning to the ground state they release the absorbed energy as visible light. If a trace of nitrogen is present in the tube, the emitted light is pink, if there is carbon dioxide, the light is white, and traces of mercury vapor result in blue-green light. In the early twentieth century, Daniel Moore, a former Edison employee, made use of this observation and commercialized the Moore fluorescent lamp.

The noble gases were invisible, but Ramsay found that when sealed in a Geissler tube and energized, neon produced a bright orange-red light that he excitedly described in his acceptance speech for the 1904 Nobel Prize, awarded for the discovery of the noble gases. Commercial application, however, would have to wait until neon could be produced on a large scale. This is when Georges Claude, dubbed the "French Edison," enters the picture. His goal was to liquify air so that it could then be fractionally distilled to yield oxygen, which was needed for the manufacture of steel. Claude managed to do this on an industrial scale, which also meant that the by-product noble gases, especially neon, could now be produced in significant amounts. Inspired by Moore's fluorescent tubes, Claude was able to produce neon tubes. He first exhibited these at the Paris Motor Show in 1910, and in 1912 installed the first-ever commercial neon sign in a Paris barber shop, opening the way for a blitzkrieg of neon signs in cities around the world. Times Square in New York became a hotbed of neon extravaganza with signs that gave the illusion of movement by cleverly turning variously shaped neon tubes on and off. Most of these signs have now been replaced by giant television screens and more efficient, eco-friendly LED lighting.

With the dawning of the computer age, neon donned another mantle. Tiny neon tubes found application as binary switches in digital circuits, and the first electronic desktop calculators had large neon-lit readouts. These now are historic relics, with neon switches and readouts being replaced by semiconductor chips and LED displays. But that does not mean neon has been dismissed. Quite the opposite. The gas is a critical component of lasers that are used in the manufacture of computer chips. Such an "excimer" laser depends on the reaction of argon with fluorine to produce a transient molecule of argon fluoride, which then relaxes back to argon and fluorine with the emission of ultraviolet light that is bounced back and forth between mirrors to produce a laser beam. The physics here is very complicated, but neon is needed to enhance collisions between argon and fluorine — the key to the workings of the laser.

About half of all the semiconductor-grade neon used in the world has been produced by two Ukrainian companies. This stems from Ukraine being a prime producer of wheat and steel. Wheat needs ammonia-based fertilizer that requires nitrogen for its production, and steel-making needs oxygen. Both of these are produced from liquid air, with neon being a by-product. With the 2022 war between Russia and Ukraine, the two companies stopped production, sending the semiconductor chip industry into a frenzy.

William Ramsay went on to identify another noble gas, radon. At the time, the danger of working with radioactive substances was not clear, and there is little doubt the nasal cancer that ended his life was caused by emissions from radon. While Georges Claude deserves credit for his work with neon, another aspect of his life was less than glowing. He publicly supported French collaboration with the Nazis, for which he was tried after the war and sentenced to life imprisonment.

Perhaps now you can look at the Five Roses sign with greater insight and admiration.

THE THIRD MAN

No visit to Vienna is complete without a ride on the Wiener Riesenrad, the giant Ferris wheel that since its construction in 1897 has been a landmark in the city's famed amusement park, the Prater. I first heard about this attraction from my father, who had been sent by my grandparents to an accounting school in Vienna in the early 1930s. He would tell me stories about the delights of Sacher cake, wiener schnitzel, and riding in the cabins of the giant wheel. After we had escaped from the Russians to Austria during the Hungarian Revolution of 1956 by crawling through the mud under the barbed wire fencing that marked the border, we ended up in Vienna. What a thrill it was to be taken to the Prater! The Riesenrad was even bigger than I had imagined.

I think one of the reasons that I became a fan of *The Third Man*, the classic 1949 film, is that a pivotal confrontation between Holly Martins, the good guy, and Harry Lime, the devious racketeer, takes place inside the cabin of the giant wheel. But there is another reason the film appeals. It has a fascinating scientific connection. The fictional racket in which Lime is involved, the selling of fake penicillin, was actually inspired by factual events. Not only that, the marketing of counterfeit drugs has since become even more real, blossoming into a global industry that leaves a swath of misery in its wake.

Penicillin was the world's first legitimate wonder drug, effective in the treatment of various microbial infections, including venereal disease. Although it first became available in 1941, supplies were limited and were mostly restricted to use by the military. This led to a large black-market trade and in 1946, the year in which *The Third Man* takes place, seven men and three women were arrested in Berlin, charged with the manufacture and sale of fake penicillin. The ring included an American doctor, two former GIs, and a former German army private who was the leader and organized the theft of used penicillin bottles. These were then filled with the antimalarial drug quinacrine and some face powder dissolved in a glucose solution. Besides being totally ineffective, the fake penicillin was also contaminated by impurities that at least in one case made a Russian officer who had been injected extremely sick.

Graham Greene, who wrote the screenplay for the movie, worked for the British Secret Intelligence Service during the war and was very aware of the illegal penicillin trade. He even knew the drug had been used as an espionage tool by Major Peter Chambers, an American intelligence officer who extracted secrets from Soviet soldiers in return for penicillin to treat their gonorrhea and syphilis. The soldiers were quite keen on this deal, given that contracting the disease could lead to a court martial. According to historians Paul Newton and Brigitte Timmermann, who described this scheme in the *British Medical Journal*, Chambers had given it the memorable codename "Operation Claptrap."

In the film, Lime organizes the theft of real penicillin from a military hospital by an orderly and enlists a physician, Dr. Winkel, to sell it to a hospital where it is to be used for treating children afflicted with bacterial meningitis. The doctor's apartment is filled with expensive objets d'art, so he obviously has been well paid for his black-market crimes. And crimes they were, since the children treated with the fake drugs were dying.

Harry Lime had hatched a clever scheme to evade the authorities by faking his own death. Holly Martins, who had come to Vienna on the invitation of his old friend Harry only to find he had just been killed in an automobile accident, becomes suspicious of the death when he hears from a witness that contrary to the police report stating that two men had carried Harry away after the accident, there was a "third man." He eventually discovers that Harry is still alive and the man killed in the accident and buried in his place was the orderly who had stolen the penicillin. Was Harry the "third man" at the scene who pushed the orderly into the path of the car?

When Holly finds out from the police investigator that Harry was involved in the penicillin scheme and is then invited to the hospital to see for himself the "murder" of the children, he agrees to cooperate with the police and trap Harry. A chase through the sewers of Vienna follows, and in the end Harry learns that crime does not pay.

Unfortunately, today, crime does pay for a lot of criminals around the world involved in counterfeit drug production. About a third of all prescription drugs in developing countries are fake, and countless numbers in the developed world are exposed to fakes from rogue online pharmacies. Some of the drugs lack any active ingredient, some are degraded medications, and some are legitimate drugs diluted to ineffective levels. Obviously, drugs for malaria, cancer, hypertension, and infections that have no active ingredient are directly responsible for killing people, but the diluted antibiotics or malaria medications are harmful in yet another way. They don't have enough of an active ingredient to treat the disease, but it is enough to foster resistance in bacteria and in the parasite that causes malaria. COVID has also give rise to fake drugs, including hydroxychloroquine and ivermectin. Even real hydroxychloroquine is a problem, because while it is ineffective against COVID, it can cause resistance against the malaria parasite.

Like Harry Lime, the fake drug producers do not care about lives. When Harry is looking down from the Riesenrad he asks Holly if he would feel pity if one of the "dots," the people walking far below, stopped moving forever if he were offered £20,000 for every dot that stopped. "Would you really, old man, tell me to keep my money, or would you calculate how many dots you could afford to spare?" Counterfeit criminals do not care how many dots they wipe off the face of the earth.

RUTHERFORD'S TRANSFORMATIONS

"All science is either physics or stamp collecting." That quote, repeated in many articles and books, is attributed to Ernest Rutherford, widely recognized as the father of nuclear physics. However, there is little evidence he ever said it. The first reference can be traced back to a book by physicist John Bernal, written in 1939, two years after Rutherford's death, in which the author makes a casual comment that "Rutherford used to divide science into physics and stamp collecting."

There is also the question of what Rutherford meant if he did indeed utter such an opinion. The quote is often interpreted as denigrating other sciences and suggesting that physics is the only legitimate pursuit. Since his own Nobel Prize, awarded in 1908, was in chemistry, and since his main collaborator at McGill University in the radioactivity experiments for which he received the prize was Frederick Soddy, a chemist, it is unlikely that Rutherford would have demeaned other sciences.

Rutherford is known to have been highly critical of theoreticians, such as Werner Heisenberg, disdaining theories he believed did not come from experiments, and it is possible that the quote may have been a dig at theoreticians. Rutherford was above all an experimental physicist and he may have used the term "physics" in the quote to mean experimentation.

The New Zealand–born Rutherford was perhaps the greatest experimentalist since Michael Faraday. He identified radioactivity as the spontaneous disintegration of unstable atoms with the emission of energy and smaller particles. Those smaller particles were identified as electrons and helium nuclei, which he named "beta" and "alpha" radiation, respectively. Disintegration could also be accompanied by energetic electromagnetic waves he called "gamma rays." Rutherford's determination of "half-lives," the time taken for half the atoms in a sample of a radioactive element to decay, laid the basis for radiocarbon dating. All this was the work carried out at McGill for which he received the Nobel Prize, but Rutherford's most famous discovery came after he left McGill to take up a post at the University of Manchester.

It was here that in collaboration with Hans Geiger and Ernest Marsden, Rutherford carried out the "gold foil experiment" that would establish the structure of the atom as we know it. As early as the fifth century BC, Greek philosophers Leucippus and Democritus had introduced the idea that all matter is composed of uniform, solid, hard, incompressible, and indestructible "atomos," from the Greek for "indivisible." Unfortunately, Aristotle, the most influential Greek philosopher, did not believe in atoms and so the concept lay dormant until the atomic theory was resuscitated by John Dalton in the early years of the nineteenth century. A hundred years later, J.J. Thomson, under whom Rutherford had studied, discovered the electron and formulated the "plum pudding model" that described atoms as uniform spheres of positively charged matter in which electrons were embedded like plums in a pudding.

Then along came Rutherford's classic experiment, which is described in every introductory chemistry and physics text. A thin piece of gold foil was exposed to a barrage of alpha particles emitted by the radioactive decay of radon, a gas that Rutherford had previously identified as a new element. A phosphorescent screen behind the gold foil flashed wherever it was struck by an alpha particle. Most of the alpha particles passed right through the gold foil, but some bounced back! An astounded Rutherford compared this to a bullet fired at a piece of tissue paper rebounding. This could only happen, he deduced, if the Thomson model was wrong and gold atoms were mostly empty space, except for a tiny, dense, positively charged mass that repelled the positively charged alpha particles. That mass came to be called the "nucleus" and was surrounded by empty space through which electrons circulated. Rutherford's theory of atomic structure was essentially correct, although further refinements have shown that electrons do not circulate randomly about the nucleus but rather are restricted to certain energy levels.

In 1971, New Zealand issued two stamps to commemorate the 100th anniversary of Rutherford's birth. One of these accurately depicts the gold foil experiment, but the other shows a reaction in which nitrogen combines with an alpha particle to form an atom of oxygen and one of hydrogen. This is an illustration of the first-ever transmutation of an element, with nitrogen being converted into oxygen, that earned Rutherford a reputation as the "first successful alchemist." Unfortunately, that reaction should not have appeared on the stamp for the simple reason that Rutherford never carried it out!

Rutherford did target nitrogen atoms with energetic alpha particles and showed that a proton, which is a simple hydrogen nucleus, was emitted, but he did not identify the other products of the reaction. Actually, he believed the bombardment had caused the nitrogen atoms to break apart into other atoms he was unable to identify. It was Patrick Blackett, working in Rutherford's lab at Cambridge, who studied this reaction extensively and in 1925 interpreted it correctly. The alpha particles had not caused the nitrogen nucleus to break apart, but rather had combined with it to form an atom of oxygen. Key to this work was Blackett's development of the Wilson cloud chamber, a device that tracks the path of charged particles, for which he was awarded the 1948 Nobel Prize in physics.

Rutherford has received many richly deserved accolades for his elucidation of radioactivity and formulation of the nuclear theory of the atom. But calling him "the world's first successful alchemist," as many publications have done, is not correct, and Rutherford would surely have opposed the design of the stamp that shows him transforming nitrogen into oxygen. While he did not carry out this transformation, he did comment on another one. In his acceptance speech for the Nobel Prize in Chemistry, Rutherford quipped, "I have observed many transformations while working with radioactive materials, but none as rapid as my own from physicist to chemist."

SCIENCE IN THE MOVIES

Any presentation on the history of medicine will include pivotal moments such as the introduction of anesthesia by William Morton in 1846 and the discovery of Salvarsan, the first truly effective antimicrobial agent, by Paul Ehrlich in 1909. Given the impact of these medical breakthroughs and their sometimes controversial history, it comes as no surprise that the discoveries caught the attention of filmmakers.

The Great Moment (1944) and *Dr. Ehrlich's Magic Bullet* (1940) tackle the epic stories in an entertaining fashion and get the science mostly right. I also like the portrayal of scientists as having lives outside the lab and the narrative that discoveries do not come about from single eureka moments but rather from a mix of capitalization on previous work by others, fruitful collaborations, and often a dose of luck.

The Great Moment is the story of dentist William Morton's discovery of ether as an anesthetic told in a series of flashbacks as Eben Frost, the first patient from whom Morton extracted a tooth using ether, reminisces about the experience with Morton's widow. As the story evolves, we learn that Morton is interested in relieving pain, as a dentist would be, and becomes captivated by colleague Horace Wells's somewhat successful use of nitrous oxide, or "laughing gas." However, he is also aware of Wells's failed attempt to demonstrate the gas to surgeon John Collins Warren and colleagues at Massachusetts General in Boston because he had not allowed enough time for the nitrous oxide to be properly absorbed.

Morton wonders if some other substance might be more reliable and seeks advice from Harvard chemistry professor Charles Jackson, who unfortunately is portrayed as somewhat of a mad scientist. Jackson tells Morton that he has had some experience with ethyl chloride as a numbing agent, but Morton mistakenly purchases a bottle of ether that he leaves on a table near the fireplace. As he is thumbing through a chemistry text, the bottle's cork pops and he inhales the ether fumes and falls asleep. This gives him the idea of using ether as an anesthetic. There is considerable poetic license taken here. Jackson actually recommended the use of ether, and Morton being accidentally overcome is fiction.

In any case, Morton proceeds to extract a tooth from Eben Frost without any sign of pain. This prompts him to approach surgeon Warren, declaring that he has something better than nitrous oxide. Although Warren, remembering the Wells fiasco, is skeptical, he agrees. But Morton had already filed a patent for an anesthetic as "Letheon," without declaring its identity. The local Medical Society finds the use of an unknown substance unacceptable and does not allow it to be administered. When Morton finds out that Warren will then have to proceed with the amputation of a young girl's leg without any anesthesia, he relents, and reveals the Letheon is ether. That leads to a happy, albeit fictional, end to the movie. In actual fact, the pivotal moment was on October 16, 1846, when Warren successfully removed a tumor from the neck of Edward Gilbert Abbott under ether anesthesia administered by Morton.

The film accurately portrays, although in a strangely wacky fashion, the legal battle between Morton, Wells, and Jackson over recognition as the legitimate inventor of anesthesia. However, history records that four years before Morton's demonstration, Georgia physician Dr. Crawford Long had placed a towel saturated with ether over a patient's mouth and successfully removed a tumor. He went on to carry out a number of amputations under ether anesthesia but did not publish his work until 1849! Long is given a one-line passing reference in the film. The acting in *The Great Moment* is borderline comical, but the film does cast light on the fascinating history of anesthesia.

Dr. Ehrlich's Magic Bullet recounts the events leading up to the introduction of Salvarsan, originally called "606," as the first truly effective drug against syphilis. The film accurately details how Ehrlich's discovery was stimulated by the observation that certain synthetic dyes are preferentially absorbed by bacteria, making them more visible under the microscope. If some sort of toxin, such as arsenic, could be incorporated into such a dye, perhaps bacteria could be killed without harming other tissues. The only way to test this "magic bullet" theory was to try. With the help of Japanese researcher Sahachiro Hata, Ehrlich synthesizes a number of dye molecules incorporating arsenic. One of these effectively treats mice infected with syphilis, a widespread disease at the time. This compound had been coded as number "606," and it is often mistakenly reported, and the film makes this error as well, that 605 experiments were carried out before one was found to be successful. But it is certainly true that "606" went on sale as "Salvarsan" with the name being coined from "safe arsenic."

At the time the film was made, movie subjects were governed by the Motion Picture Production Code that had been introduced in 1930 and prohibited mention of "sex hygiene and venereal disease." The film's producers claimed that not addressing one of Ehrlich's major discoveries was unfair to his legacy and managed to get approval providing that treatment of patients with syphilis was not shown and that advertising of the film would not mention the disease.

There was yet another controversy. Paul Ehrlich was Jewish, and by 1940 the Nazis had expunged all references to his achievements. The U.S. had not yet entered the war and American films were popular in Germany. When screenwriter Norman Burnstine pitched his idea for a film about Ehrlich, he wanted to address anti-Semitism and reportedly said, "There isn't a man or woman alive who isn't afraid of syphilis . . . let them know that a little kike named Ehrlich tamed the scourge," adding, "and maybe they can persuade their hoodlum friends to keep their fists off Ehrlich's co-religionists." The producers decided that the film should stay clear of such issues, and the words "Jew" or "Jewish" were never mentioned. Obviously, there's more to movies than what we see on the screen.

THE BIG NICKEL

The largest coin in the world is a 1951 Canadian nickel. But you can't put this one in your pocket since it is about 30 feet high and 2 feet wide. The "Big Nickel," a popular tourist attraction in Sudbury, Ontario, commemorates the 200th anniversary of the isolation of metallic nickel by Swedish mineralogist and chemist Baron Axel Frederik Cronstedt and indirectly also pays homage to the ingenuity of another chemist, Ludwig Mond, who developed the first commercial process to produce pure nickel. The Sudbury area is rich in nickel-bearing ore and has a long history of supplying the metal to the world, a process in which Mond also played a large part.

Back in the seventeenth century, German miners searching for copper discovered a reddish ore that looked to be a source of the metal. However, try as they might, they were unable to extract any copper and concluded that a prank had been played on them by Nickel, a mischievous demon in German mythology. They named the ore that would not yield copper "kupfernickel," meaning "copper demon." It didn't yield copper for the simple reason that it didn't contain any, as was eventually shown by Cronstedt, who in 1751 heated kupfern-ickel with charcoal and produced a metal not seen before that clearly was not copper. He dropped the term "kupfer" and named the metal "nickel." Cronstedt had discovered a new element!

Nickel is shiny, sturdy, and resists corrosion, which made it ideal for use in coins and stainless steel. Armor on ships previously made of iron was greatly strengthened when the iron was alloyed with nickel. Producing pure nickel from its ore, however, was a challenge, one that was eventually met by Ludwig Mond, who certainly did not set out to purify nickel. Like many discoveries, this one came about in a round-about fashion.

Mond was born in 1839 to a prominent Jewish family in Germany. His father was able to send him to the best schools, including the University of Heidelberg, where he studied under Robert Bunsen of burner fame. He left before completing his doctorate, apparently more interested in practical than theoretical chemistry. Young Mond found a job with a factory producing acetic acid by the distillation of wood; there he found an economical way to combine acetic acid with copper to produce verdigris, a much sought-after green pigment. The next step in the chemist's career was taken at a Leblanc soda works factory that produced sodium carbonate, often abbreviated as "soda," a key chemical in making paper and glass. One problem was the large amounts of a waste product, calcium sulfide, produced along with the soda. Mond managed to develop a process to convert the calcium sulfide into marketable sulfur. This brought him to the attention of the John Hutchinson Company, an English soda manufacturer, and prompted a move to England in 1862. He went to work for the company but soon decided to set out on his own with partner John Brunner.

The major competitor of the Leblanc process was the "ammonia-soda" process developed by Belgian chemist Ernest Solvay. Not only was this more efficient, but Mond believed that the waste it produced, ammonium chloride, could be profitably converted into chlorine gas. He traveled to Belgium, where he convinced Solvay to grant a license to Brunner,

Mond, and Company to produce sodium carbonate by his method. It was during the conversion of the ammonium chloride to chlorine that Mond made the discovery that would launch him into the next step of his career, the production of nickel.

The production of chlorine involved vaporizing ammonium chloride and passing it through a network of pipes and valves. Nickel, because of its resistance to corrosion, was used to construct the valves. While the process worked well, the valves tended to leak as they became coated with a mysterious black deposit. Mond, intrigued, studied this deposit and found to his amazement that with heat it turned it into shiny, metallic nickel! Further analysis revealed that the deposit was nickel carbonyl, formed by the reaction of nickel with carbon monoxide. But where did the carbon monoxide come from? It turned out that the pipes were periodically flushed with carbon dioxide to blow out residual vapors of ammonia and the carbon dioxide was contaminated with traces of carbon monoxide.

While some may have just concentrated on fixing the problem by eliminating the carbon monoxide, Mond realized that he had made an important discovery. He had produced extremely pure nickel! Exploiting this serendipitous discovery, he founded the Mond Nickel Company and purchased nickel ore mines around Sudbury. Here the ore was smelted and impure nickel shipped to a refinery in Wales, where it was be treated with carbon monoxide to yield the nickel carbonyl that would then be heated to produce pure nickel.

Mond's soda and then his nickel business made him into a wealthy man. He was generous to his employees, being one of the first industrialists to offer paid vacations, eight-hour workdays, and fringe benefits such as recreation clubs and sports fields. A great promoter of chemical research, he donated large sums to Britain's Royal Institution, the Children's Hospital of London, and to his alma mater, the University of Heidelberg. Mond was also an enthusiastic supporter of the arts and donated his sizeable collection of Italian Renaissance paintings to the National Gallery in London, the largest single gift the museum has ever received.

While the 1951 coin on which the Big Nickel was modeled was made of 99.9 percent nickel, the giant sculpture is made of stainless steel, which of course does contain some nickel. And those nickels in your pocket now? Made of steel and copper with a thin plating of nickel. Just enough to now make you think of the contributions of Ludwig Mond to science, the arts, and social reform.

TIN PAN ALLEY

I don't know exactly what sound a tin pan makes when it is banged. That's because finding a tin pan is very difficult. It is easy to find an iron or copper pan coated with tin to protect it from corrosion, but cookware made from pure tin is rare because the metal is very soft. Back in the 1800s, though, cups and plates made from tin were around, including some frypans called "cowboy frypans" that were popular for outdoor cooking. Tin pans were also favored by prospectors. They were light and easy to carry around — very handy when panning for gold!

Why am I interested in banging a tin pan? Because I'm fascinated by "Tin Pan Alley" and how it got its name. Starting in the 1890s, a number of music publishing firms settled on a small stretch of 28th Street between 5th and 6th Avenues in New York. Although Thomas Edison had invented the phonograph in 1877, recorded music in the 1890s remained more or less a curiosity. Edison's cylinders produced about two minutes of scratchy music, a virtual miracle at the time, but the cylinders were certainly no replacement for live music.

Pianos in homes were very popular, which also meant that the marketing of sheet music became a lucrative business. Composers flooded publishers with their music, and the publisher then worked at boosting sales by hiring pianists to play new pieces to potential customers. The sounds of music being pounded out on worn-out pianos filled 28th Street. Apparently, composer and journalist Monroe Rosenfeld likened the clatter to the sound of tin pans being banged in an alley. Another account suggests that the pianists tried to make sound produced by improperly tuned pianos less "tinny" by hanging paper strips from the strings. Both these stories may be apocryphal, but there is no doubt that the strip of 28th Street where the music publishing industry was centered came to be known as Tin Pan Alley.

Where would Rosenfeld have gotten his analogy? Perhaps he was familiar with the antics of French protesters in the 1830s who opposed the regime of Louis Philippe I by parading through the streets and demonstrating their disapproval of the government by beating saucepans to make noise. Such saucepans were commonly referred to as tin pans, even though they were likely made of another metal lined with tin. More recently, in 2012 in Montreal, students and supporters banged pots and pans as they paraded to oppose a hike in tuition. When the government passed a law to limit the scope of student protests, the response was more parades with even more banging of cookware!

How about the alternate explanation of "tinny" sound for the origin of Tin Pan Alley? Since tin, being very soft, was never a good substitute for other metals in most applications, its name became associated with something that was not quite up to par. Such as the sound from a rickety piano. The pianos may have been wobbly, but not so the music churned out by Tin Pan Alley. Songs like Irving Berlin's "Alexander's Ragtime Band" and George Gershwin's *Rhapsody in Blue* became all-time classics.

The bestselling song in Tin Pan Alley history, however, was not written by a composer whose name rolls off the tongue of music lovers, but by Tin Pan Alley pioneer Charles K. Harris, who became known as the "king of the tearjerkers." That song was "After the Ball," in which a man tells his niece that he never married because he saw his sweetheart kissing another man at a ball and he refused to listen to her explanation. Many years later, after she had died, he discovered that the man was her brother. The sheet music for that song, which eventually became incorporated into the musical *A Trip to Chinatown*, sold more than five million copies! The show ran for two years on Broadway, becoming the longest running Broadway musical in history up to that time. The plot is actually reminiscent of *Hello, Dolly!*, with a widow arranging some hookups in a humorous fashion.

Tin Pan Alley withered away after record players, radio, and then television delivered music to the population in an easy fashion. However, five buildings on 28th Street have been landmarked and preserved as the Tin Pan Alley Historic District, with a plaque on the sidewalk declaring it to be "the legendary Tin Pan Alley where the business of the American popular song flourished during the first decades of the 20th century."

The role that tin played is still somewhat ambiguous, but since I couldn't find a tin pan to sound out, I did kick a tin can around. It did make a tinny sound. But these days "tin cans" are actually made of aluminum. So is "tinfoil." As a finale, let's note that when Edison uttered "Mary had a little lamb" into his phonograph, to be reproduced later, the recording was on tinfoil wrapped around a hand-cranked cylinder into which indentations had been etched with a needle attached to a diaphragm. You can listen to it online. A real piece of history. But it does sound tinny.

VALENTINE'S MEAT JUICE

Mann Valentine of Richmond, Virginia, loved his wife. He was totally distraught when she fell ill and appeared to be withering away, unable to eat solid foods. Back then in 1870, with doctors unable to offer much help, Valentine decided to take matters into his own hands. He had some rudimentary familiarity with nutritional science, which at the time embraced the notion of "muscle to muscle" that had been around since the ancient Greeks downed the muscular flesh of animals in the hope of gaining strength. "Just extract the essence of meat," was the idea that now occurred to Valentine. Perhaps meat juice would be the key to restoring his wife's strength!

The soon-to-be inventor reportedly went down to his basement and worked out a method of cooking meat and squeezing out its juice. Administering the concoction to his wife led to a remarkable improvement from which, Valentine thought, the public should benefit. Within a year he had set up a company and began producing Valentine's Meat Juice, sold in what would become an iconic pear-shaped amber bottle. Booming sales following enthusiastic testimonials from patients and physicians made Valentine a wealthy man.

The rather remarkable feature of this story is that Valentine seems to have reinvented the wheel! He was apparently unfamiliar with Liebig's Extract of Meat, which had been introduced in Europe in 1865, based upon the ideas of German chemist Justus von Liebig who at the time was one of the world's leading scientists. Liebig had discovered the presence of compounds of nitrogen in urine and postulated that these stemmed from the breakdown of muscle during activity since muscle was known to consist of nitrogen-containing proteins. He was concerned that many people could not afford to eat enough meat to sustain health and in 1847 began to experiment with developing a concentrated, affordable, nutritious meat substitute. Liebig found that soaking lean meat in a vigorously stirred dilute hydrochloric acid solution resulted in a paste that could be strained to yield a concentrated meat extract.

When Liebig published his method, druggists and physicians began to make small batches of "beef tea." In 1851, physician William Beneke reported in the *Lancet* his successful use of the tea in the treatment of tuberculosis, typhus, and "stomach derangements." Liebig concurred, championing the use of beef tea as medicine, but recognized that there was little commercial potential for the extract since the process of making it was tedious and European beef was expensive.

George Giebert, a German engineer who had built roads in Brazil, now approached Liebig with a possible solution. Lots of cattle were being raised in South America, mostly for their hides with the meat often being discarded. Labor was also cheap, and Giebert suggested buying cattle farms in South America and shipping machinery from

Europe to produce the extract. Liebig liked the idea, and in 1865 the Liebig's Extract of Meat Company was established and soon the product hit the marketplace. At first, it was sold as a remedy for "weakness and digestive disorders" but soon claims became more elaborate. Liebig himself touted its ability to allay "brain excitement," and at a British Pharmaceutical Conference, speakers asserted that "probably no food available was as effective at restoring the tissues of the sick."

Copycat products, such as Bovril, also mushroomed, with Liebig warning of imitators and urging consumers to buy only the genuine version, which was inspected by himself and featured his signature on the label. As the meat extracts increased in popularity, some scientists began to look on them with a wary eye, especially after analyses showed that Liebig's extract actually contained little protein. Then in 1868, German physiologist Edward Kemmerich published his results of an experiment in which dogs exclusively fed on the meat extract soon died. Then in 1872 physician Edward Smith declared that Liebig's extract lacked the nutrients of meat and was like "the play of *Hamlet* without the character of Hamlet." That attack paled in comparison with the rhetoric of another physician, John Milner Fothergill, who opined that "all the bloodshed caused by the warlike ambition of Napoleon is as nothing compared to the myriads of persons who have sunk into their graves from a misplaced confidence in beef tea."

In light of the fading aura of the extract as a medicine, the Liebig Company switched to promoting it as an "inexpensive source of meat flavor that sailors, explorers, soldiers, and domestic cooks could use to produce a nutritious and tasty soup by adding potatoes and vegetables to a broth made from the extract." Then came the brilliant move by introducing Liebig trading cards that came with each bottle. These were beautifully colored cards that at first depicted kitchen scenes with cooks preparing soup with ease, but then expanded into portrayals of scientists, writers, composers, and idyllic historical scenes. The cards became a phenomenon among collectors and are regarded as one of the most successful advertising campaigns in history!

The Liebig's Extract of Meat Company no longer exists, but one of its products, the Oxo bouillon cube, developed in 1911, is still around, and advertised in an ingenious fashion. In 1920, the Liebig Company purchased a building in London that featured a tower that they planned to equip with an illuminated advertising sign. When permission for this was refused, three windows on the tower were redesigned to be shaped like the letters "o" and "x" to spell out "OXO"!

As far as Mrs. Valentine goes, she passed away just two years after her husband introduced his meat juice. But the profits from the product were enough to allow him to indulge in his passion for collecting artifacts that were eventually displayed in the Valentine Museum in Richmond. Founded in 1898, the "museum that meat juice built," has become a major attraction with exhibits depicting the city's rich history.

THE SHOT TOWER

Education sometimes takes a circuitous route. Like my learning about the fascinating history of "shot towers." Appropriately enough, it all began with Sherlock Holmes, investigator extraordinaire. I was trying to track down Sidney Paget's original illustrations as they appeared in the story "The Red-Headed League," published in the *Strand Magazine* in 1891. As one might expect, a reproduction of the issue wasn't difficult to find online, replete with Paget's wonderful drawings. Glancing at the last page of the Holmes story, my eye caught the next article in the magazine, entitled "Up the Shot Tower," accompanied by a picture that looked something like a lighthouse. And thus began a journey through history that would end with an awe-filled upward gaze on Dominion Street in Montreal!

"Shot Tower" turns out to be appropriately named, given that the structure is designed to manufacture "shot," tiny lead pellets fired from shotguns. Propelling those pellets requires gunpowder, a mixture of sulfur, charcoal, and saltpeter (potassium nitrate), first described in China in the ninth century. Likely an accidental by-product of experiments seeking to find the "elixir of life," as suggested by the substance's Chinese name that translates as "fire medicine." By the twelfth century, the Chinese had developed "fire lances," essentially bamboo tubes packed with gunpowder that would forcefully eject arrows or bits of metal upon ignition.

Before long, Chinese metalworkers developed real cannons cast of brass or iron, and by the fourteenth century, the technology had landed in Europe with craftsmen designing "hand cannons." These, loaded with lead pellets, were the forerunners of all guns. The use of lead was already well established, with the smelting of the metal's ore, galena (lead sulfide), dating back to around 6000 BC! Lead is easy to cast and shape, a property with which the Romans were very familiar. They famously constructed water pipes and made dining vessels from lead, oblivious of the metal's toxicity. But pouring lead into wooden molds to make "shot" in the amounts required for weapons was an inefficient process. Then in 1782 came the break-through: the shot tower!

English plumber William Watts gets credit for the discovery. As a plumber, he knew all about working with lead. Indeed, the name of the profession derives from "plumbum," the Roman term for the metal, which also explains why Pb is the chemical symbol for the element.

There are a number of stories circulating about how Watts hit upon the idea of a shot tower. According to one, after a day's work of casting lead for shot, he imbibed of ale a bit too freely, fell asleep, and had a dream in which rain turned to lead, covering the wet ground with tiny pellets. Intrigued by this vision, he melted some lead and dropped bits from different heights, noting that the drops became rounded as they fell. Another account suggests that Watts knew that castles were sometimes defended by pouring molten lead on the enemy, with the metal ending up in the moat, from which it was recovered in the shape of balls. The most likely explanation, though, for sparking the idea of a shot tower is Watts's observation that as a drop of water falls from a faucet, its shape changes from a teardrop to a sphere. Perhaps drops of lead would do the same!

But how do you keep the molten droplets from splattering when they hit the floor? How about allowing them to fall into cold water? Height turned out to be an issue with drops of lead having to fall a greater distance than drops of water to achieve a spherical shape. First, Watts knocked out a hole in the ceiling so he could drop the lead from the second floor of his house, undoubtedly to the dismay of Mrs. Watts. When this wasn't high enough, he added stories until the house became a tower! Pouring molten lead from the top through a copper sieve now resulted in perfect shot being recovered from the tub of water placed at the bottom.

Watts's "shot" was soon heard around the world and "shot towers" began to sprout up everywhere. One of these was on the banks of the Thames in London, as described by the author of the article I had come across in the *Strand*. That tower had 327 steps and produced a silvery rain of millions of pellets every day. The first tower in America, Philadelphia's Sparks Shot Tower, was built in 1808 and produced tons of ammunition during the war of 1812 and the Civil War. At 234 feet, Baltimore's Phoenix Shot Tower, dating back to 1828, was the tallest structure in the United States until the Washington Monument was completed in 1884.

A number of shot towers can still be found around the world, although none are active, the technology having been replaced by modern machinery. Unfortunately, both Watts's original and the London tower that set me on my journey are gone, as I discovered when searching for still-existing towers. But it was during that search that I suddenly gasped when I came across a mention of Montreal! I was astounded to discover that we have a shot tower right here. Of course I had to go see it!

The Stelco Tower looks to be about ten stories high and can be found on Dominion Street, right by the Lachine Canal, just east of the Atwater Market. I must have seen it numerous times as I bicycled along the canal but never noted it. Now I gawk at this wonderful historic relic every time, as will you if you have a chance to wander that way. And there you have the shot tower story, lock, stock, and barrel! Thanks, Sherlock.

ACONITE MURDER

It was a real puzzler. On autopsy, Dr. Ohno could find no evidence of any prior condition to explain why the thirty-three-year-old Japanese woman died of an apparent heart attack that day in 1986. While out with her friends, she was overcome by nausea and complained of a loss of feeling in her extremities. When she vomited violently, her friends called for help, but in the ambulance the unfortunate lady developed an irregular heartbeat that could not be corrected by defibrillation and she died. Since the death could not be readily explained, the police were informed.

Upon questioning, the victim's husband revealed that he had been married twice before and both wives had died, one of a heart attack the other of myocarditis. That was curious, and suspicions were raised further when it was discovered that the husband had recently insured his wife's life for a staggeringly large sum. Dr. Ohno now suspected poisoning, and his thoughts turned to aconite, a rapidly acting toxin known to cause ventricular fibrillation and paralysis.

Aconitum napellus is a perennial herb commonly known as monkshood since its purple flowers resemble the hoods worn by monks. All parts of the plant contain extremely toxic alkaloids, with aconitine leading the pack. Swallowing just 2 milligrams of these compounds or 1 gram of the root can be fatal! The toxicity of a crude extract of the plant, known as "aconite," was already known to the Romans, who used it as a method of execution. We know that Shakespeare was aware of its toxicity, specifically mentioning Aconitum in *Henry IV*. It is likely also the poison he had in mind for a lovestruck Romeo's suicide.

While aconite can account for paralysis and irregular heartbeat, there were a couple of problems with the poisoning theory. There was no way at the time to test blood or tissue samples for the tiny amounts of alkaloids that could have caused death. And there was also an issue with the time frame. The last time husband and wife were together was about an hour and a half before her collapse, much too long for the aconite's effects to be manifested.

Luckily, the police decided to keep some blood samples, and just nine months later that foresight paid off. A method of detecting very small amounts of Aconitum alkaloids using the combined techniques of gas chromatography and mass spectrometry was worked out, and they were indeed detected in the stored samples. Still, this was not enough to connect the husband to the poisoning because of the time problem. But the police kept digging, and four years later discovered that the husband had purchased a bunch of *Aconitum napellus* plants! That led to his arrest and indictment for murder.

The investigation revealed something else as well. The accused had also purchased some puffer fish, a delicacy in Japan. Since the species harbors the potent poison tetrodotoxin, Japanese chefs are specially trained to remove the toxic organs before serving fugu, as the dish is known. When authorities tested the victim's stored blood, they found tetrodotoxin!

Now Dr. Ohno had an idea. He knew that the toxic effects of aconitine are due to enhancing the influx of sodium ions into nerve cells, and that tetrodotoxin kills by starving nerve cells of sodium. Could the tetrodotoxin delay the action of aconitine? Could these toxins act in an antagonistic fashion? A series of experiments on mice showed that the toxic effects of aconitine were markedly reduced by the oral co-administration of tetrodotoxin. Based on the evidence presented, the husband was found guilty of murder and sentenced to life in prison. While he admitted that he was interested in chemistry and had bought the plants and the puffer fish to experiment with, he maintained that he had nothing to do with his wife's demise.

Was he a clever chemist who had found a way to delay the action of aconitine and deflect suspicion from himself, or was he trying to mix the two potent toxins to ensure a quick death but just happened to accidentally postpone it? We will never know, but his murderous plan did shed light on the combined effects of aconitine and tetrodotoxin.

While there have been many other criminal poisonings with aconite, including a widely publicized 2010 case in England in which a scorned Mrs. Lakhvir Singh murdered the lover she had been having an extramarital affair with, there have also been a number of accidental poisonings. In Traditional Chinese Medicine, aconite is used, among other conditions, to treat faint pulse, impotence, and "Yang deficiency." If not properly detoxified by toasting or steaming, such aconite preparations can be very dangerous. Aconite is also used in homeopathy to treat conditions ranging from respiratory infections and toothache to vertigo and kidney stones. Since according to the tenets of homeopathy these preparations are diluted to the extent that they only contain the "memory" of aconite, they are harmless. They also lack any evidence of efficacy.

Finally, we note that the ancient Greeks used juice from monkshood to poison arrows, and historically aconite was also used to kill carnivores such as panthers and wolves with poisoned bait. Hence the alternate name for the herb, "wolfbane." There is yet one other connection. Should you be worried about werewolves, you might consider keeping a few sprigs of wolfbane around. They are said to keep lycanthropes at bay.

THE PITFALLS OF PROPOSITION 65

I know very little about Pilates, and I must admit I had never heard of Cardi B. It turns out that the two names have something in common. They were both mentioned in questions directed my way about chemical risks. "Should I return my Pilates ring?" one correspondent inquired, while another wanted to know why Cardi B was promoting toxic clothing. Both queries came with attachments of pictures of warning labels. The Pilates ring warned that "this product can expose you to chemicals including lead, which is known to the State of California to cause cancer and birth defects or other reproductive harm," while the question about the toxic clothes was accompanied by a photo of a similar label on a brightly colored bikini stating, "This product can expose you to Di(2-ethylhexyl) phthalate, lead and cadmium, which are known to the State of California to cause cancer, birth defects, and other reproductive harm."

Joseph Pilates was a sickly child born in 1883 in Germany. After reading about the ancient Greeks' emphasis on athletics, he turned to exercise as a possible remedy for his asthma. It seems to have worked because he became an avid skier, boxer, and gymnast. In 1912, Pilates immigrated to England, where he was interned as an "enemy alien" when World War I broke out. It was during his confinement that Joseph developed a system using bed springs as exercise equipment for his fellow internees. While the 1918 influenza pandemic devastated the country, it seems none of Pilates's trainees died, an outcome for which he took credit. After the war he returned to Germany, where his exercise regimen was adopted by dancers. In 1926, Pilates immigrated to the U.S. and launched his mind-body system of exercise, achieving widespread popularity. Resistance training, such as compressing a spring, is a big part of the program, and a Pilates ring made of plastic or plastic-coated metal provides such resistance. It is the plastic component that is responsible for the warning label.

Cardi B, as I now have learned, is a popular rapper whose sexy clothes are part of her image. She has collaborated with Fashion Nova to produce a line of apparel that includes various items made of vinyl. These are the ones that cause concern when customers discover the warning label after purchasing the garment. So, is working out with a Pilates ring or wearing vinyl pants à la Cardi B a health risk? It all depends on how one interprets a controversial California law originally known as the Safe Drinking Water and Toxic Enforcement Act that was passed in 1986.

Let's take a quick trip back to the days before Proposition 65, as the bill was described on voters' ballots, became law. In the early 1980s, the quality of drinking water in California came under scrutiny as problems of nitrate runoff from fertilizer and contamination by solvents from the fledgling silica chip industry dominated headlines. The state government proposed a law that would prevent businesses from discharging substances known to be toxic into water systems and would also require warnings on items that contained substances believed to pose a risk of cancer or reproductive harm to the consumer.

The proposed law became a hot potato, with industry fiercely opposing it and environmental groups lobbying for its passage. A group of Hollywood celebrities led by Jane Fonda, Whoopi Goldberg, Rob Lowe, and Michael J. Fox toured California in buses dubbed the "clean water caravan" to boost support for Proposition 65. The public, obviously thirsty for clean water, voted overwhelmingly, 63 percent to 37 percent in favor of the bill.

Proposition 65 quickly proved to be effective when it came to regulating discharges by industry. Drinking water quality improved. However, the proposed warning labels on consumer products opened up the proverbial can of worms. The "dose makes the poison" is of course the cornerstone of toxicology, but with carcinogens and hormone-disrupting chemicals the establishment of a safe dose is a challenge. For carcinogens, a somewhat arbitrary danger level was set at a dose that would present more than a 1 in 100,000 risk of cancer assuming a lifetime exposure. Reproductive toxins were deemed to be dangerous at a dose in excess of 1/1,000th of the no observable adverse effect level (NOAEL) as determined by animal studies.

There are huge safety factors built into both these categories. Lifetime exposure makes sense for some substances in drinking water, since it is indeed regularly consumed over a lifetime. It makes much less sense for lead in a Pilates ring. There is no question that lead and its compounds are highly toxic, and it is even conceivable that trace amounts may leach out when handling the equipment's polyvinyl chloride (PVC) coating. With a stretch of credulity, it is even possible that this would constitute a lifetime risk, if it were handled in this fashion over a lifetime. But that is hardly the case.

There is another point to be made. Lead compounds were once widely used in PVC as a stabilizer. When this plastic degrades, it releases hydrochloric acid that then further catalyzes the degradation. Lead compounds neutralize this acid and act as a stabilizer. However, when the toxicity of lead became widely known, these were replaced in PVC by other stabilizers. Unless the plastic comes from old recycled PVC, which is very unlikely because this plastic is not recycled, the probability is that Pilates rings do not contain lead at all. Why then the warning? Basically, as protection from profiteering lawyers who have built careers on reaching out of court settlements by threatening companies with lawsuits for not being compliant with Proposition 65. It is easier to affix the warning than to go through the complexity and expense of a trial where you have to prove your innocence. Since businesses cannot know which of their goods may end up being purchased by Californians, to avoid possible legal action, they affix a Proposition 65 label on any item requiring it no matter where it is sold.

Now for the Cardi B–inspired vinyl clothing. To make vinyl soft and pliable, plasticizers such as phthalates are added. Indeed, these have the potential for reproductive toxicity. Again, amounts are important. The NOAEL as determined in animal studies is the maximum amount that can be administered without causing any effect. Since humans are not large rodents, an added safety factor of 1,000 is built in. Should one worry about wearing pants that might contain phthalates that may slightly exceed this remarkable safety margin? I think not, even if one were going to dine on such apparel. One might, however, argue that once such garments are discarded, whatever toxins they contain can find their way into the environment and that justifies a warning label.

While Proposition 65 has certainly curbed the release of toxic substances into the environment by industry, having warning labels on items ranging from hammer handles to fishing lures may amount to crying wolf, as in the classic Aesop's fable. Warnings about a real wolf coming to the door may then go unheeded.

RED LIGHT THERAPY

As the story goes, back in seventeenth century Amsterdam, ladies of the night carried red lanterns to signal sailors that they were available. Supposedly the red light had another effect. It camouflaged the skin imperfections these women often had. In light of subsequent research, maybe these lanterns did more than just hide scars and lesions, maybe they actually helped heal them!

A wall engraving dating back some 3,000 years shows Queen Nefertiti and her children soaking up the rays of the sun, suggesting that the ancient Egyptians believed in the health benefits of exposure to sunlight. The Greeks and Romans were fond of solariums, and in preparation for the original Olympic Games, Greek athletes were encouraged to enhance their strength by exposure to sunlight for several months. However, it was not until the late nineteenth century that the spotlight of science began to shine on the therapeutic effects of light.

Dr. Niels Ryberg Finsen, born in the Faroe Islands and educated in Denmark, suffered from Niemann-Pick disease, a rare ailment in which harmful amounts of fat accumulate in internal organs and dark areas of pigmentation mar the skin. His belief that sunlight might help his condition set the stage for a research career focusing on the possible healing properties of light. Seeing that sunlight had no effect on his own condition, Dr. Finsen switched to exploring artificial light and began a collaboration with the Copenhagen Electric Light Works to produce an electrical carbon-arc lamp, a project that in

1895 led to a lucky accident. Niels Mogensen, an engineer with whom Dr. Finsen was working, suffered from lupus vulgaris, a skin infection characterized by terrible disfiguring lesions caused by Mycobacterium tuberculosis bacteria. No treatment Mogensen had tried worked, but while working on the carbon-arc lamp he noticed the lesions improved. The engineer became Dr. Finsen's first patient, and after only a few days of treatment with the Finsen light his condition resolved.

The doctor then went on to try his lamp on patients with smallpox scars with highly satisfactory results. While his lamp produced full spectrum light, Dr. Finsen proposed that it was the red end of the spectrum that had the healing effect. When presented with this theory, the chief physician at a Copenhagen hospital rejected it out of hand. Finsen retorted, "You might at least try not to laugh at me." The laughter stopped when doctors in Norway reported newly diagnosed smallpox patients sequestered in "red rooms" recovered without ever developing scars. There was now enough evidence to convince the mayor of Copenhagen, with backing from a number of donors, to establish the Medical Light Institute with Finsen as its director. The results of treatment with Finsen light were impressive. Of patients afflicted with lupus vulgaris, 83 percent were cured! Within a few years, forty Finsen Institutes were established in Europe and America. Treatment of lupus vulgaris with Finsen lamps continued until antibiotics were introduced half a century later.

In 1903, Dr. Finsen received the ultimate recognition, the Nobel Prize in Medicine for launching the field of phototherapy. Unfortunately, by this time he was confined to a wheelchair and was unable to travel to Stockholm to receive the prize and died just a year later.

Now let's skip ahead to 1967, when Hungarian physician Endre Mester tried to repeat an experiment by American Paul McGuff, who had used a red laser beam to destroy a cancerous tumor implanted into a laboratory rat. Unbeknownst to him, Mester's laser was much weaker and had no effect on the tumor, but to his surprise caused rapid healing of the wound where the tumor had been implanted! Furthermore, the light stimulated the regrowth of hair at the site! Mester coined the term "photobiostimulation" for this low-level red light therapy.

Since then numerous researchers have explored the potential of red light therapy. That includes NASA scientists who found that red light boosts plant growth on the International Space Station and allows astronauts' injuries to heal faster. Other studies have shown possible benefits in pain relief, acne treatment, blood circulation, asthma, inflammatory conditions, stroke, and even hair growth. Light-emitting diodes implanted in special helmets, or attached directly to the forehead, have shown some tantalizing benefits in depression, Parkinson's disease, and cognitive enhancement. Finally, the effect of light on COVID has been studied. The violet/blue end of the spectrum has been shown to inactivate bacteria and some viruses, and at least in experimental animals, red and near-infrared light reduce respiratory disorders similar to complications associated with coronavirus infection.

The mechanism of action of red light has been explored with the prevailing theory being that damaged cells produce nitric oxide, which binds to and inactivates cytochrome oxidase, an enzyme that is required to produce adenosine triphosphate (ATP), the molecule that releases energy to fuel cellular processes. Light in the red (600 to 700 nanometer wavelength) and near-infrared (760 to 940 nanometers) regions of the spectrum liberates nitric oxide from the enzyme, allowing more ATP to be produced, normalizing cell function. Furthermore, the nitric oxide released is a messenger molecule that has beneficial effects on the immune system, on blood vessel dilation, and on blood clotting.

As is often the case, inventive marketers and pseudo experts are prone to hyping research results beyond what the data actually shows. According to some promoters, red light therapy is a cure for whatever ails you. Such claims should raise the red flag of alarm.

These days, lanterns in red light districts have been replaced with red neon signs. But that is not the place to go for red light therapy. A better bet would be the commercially available red LED arrays that are at least supported by some, albeit not overwhelming, evidence.

THE LEIDENFROST EFFECT

"Bertie," as Queen Victoria and Prince Albert's eldest son was known to the family, was not a stellar student, much to the dismay of his parents. They tried to groom the Prince of Wales for his future role as king by sending him on tours of educational institutions, hoping to spark a flame of interest. In 1859, the eighteen-year-old Bertie was introduced to Professor Lyon Playfair, who was to be his guide for a tour of the chemistry labs at Edinburgh University. Playfair thought the prince would enjoy one off his favorite demonstrations and proceeded to pour some molten lead over the fingers of his assistant. Indeed, Bertie was amazed that the man's hand was not scalded and asked if he could try the experiment. Professor Playfair instructed him to dip his hand in water, shake off the excess, and extend the royal fingers as he proceeded to pour the molten lead over them. The hand that would rule England as King Edward VII from 1901 to 1910 was unharmed. Young Bertie had learned about the Leidenfrost effect!

After graduating as a physician in 1741, Johann Gottlob Leidenfrost went on to teach medicine, physics, and chemistry at the University of Duisburg in Germany. He published a number of scientific papers, including "A Tract About Some Qualities of Common Water" in which he described the effect that would be named after him. He had observed that droplets of a liquid placed on a surface considerably hotter than its boiling point skittered across the surface before turning into a gas instead of instantly evaporating. At a lower temperature, but still above the boiling point, the droplets immediately vaporized.

In a classic experiment, Leidenfrost placed a drop of water on a polished iron spoon heated over glowing coals and recorded that the drop "does not adhere to the spoon, as water is accustomed to do, when touching colder iron." Rather, it hovered over the spoon as a spherical globule for close to half a minute before evaporating. Upon placing a candle behind the drop, Leidenfrost observed that light passed between the spoon and the liquid, revealing the presence of a thin layer of vapor on which the droplet floated. This effect is often put to a practical use by cooks wishing to know if their frying pan is hot enough before adding food. If the metal has been heated enough, a few droplets of water sprinkled over the surface will skitter about before evaporating while a lower temperature will result in the water turning into steam as soon as it touches the pan.

The molten lead "trick" makes use of the same principle. Contact of water with the hot lead forms an insulating layer that prevents a burn. But as Adam and Jamie demonstrated on *Mythbusters*, the molten lead has to reach a temperature of 450°C, some 120 degrees above lead's boiling point, to prevent injury. They cleverly showed this by first testing on sausages instead of fingers before successfully plunging their digits into the molten lead. Thanks to the Leidenfrost effect, we have the curious phenomenon of a higher temperature causing less harm than a lower one.

The oldest lie detector system in the world also relies on the same effect. In an ancient ritual used by Bedouin tribes, a member who professes innocence after being charged with a crime is given the chance to prove that he is not guilty. All he has to do is lick a red-hot spoon without blistering his tongue! The idea is that if he is guilty and nervous, his mouth will be dry and the tongue will burn as soon as it touches the hot metal. If he is innocent, there will be enough saliva in the mouth to offer protection via the Leidenfrost effect. This "trial by fire" is known as Bisha'h and has been banned by all governments in the Middle East except apparently Egypt, where it is still performed. A good liar can probably muster up enough protective saliva, and I suspect there are some innocents who wish they had not agreed to lick the spoon.

It would be great to report that the dramatic demonstration of the Leidenfrost effect triggered an interest in chemistry for the Prince of Wales, but such was not the case. It seems the young man was more interested in debauchery. When his exasperated parents sent him to a military training camp in Ireland to correct his rambunctious behavior,

Bertie contrived to smuggle actress Nellie Clifden into his quarters to initiate him into manhood. When his mother, the Queen, heard about this, she sent her husband, Prince Albert, to scold him. Albert reputedly told Bertie, "I knew that you were thoughtless and weak, but I could not think you depraved." Albert died just two weeks after this chewing out of his son, and Victoria blamed Bertie for precipitating the death by causing stress to his father. She never forgave him. "I never can, or shall, look at him without a shudder," she said, and never did.

Bertie was not totally disinterested in science. He was certainly interested in nutrition. Over-nutrition. He consumed gigantic meals and became, let's just say, portly. His hunger, though, was not only for food. He had a spectacular appetite for sex! As time went on and his belly grew, he had a difficulty positioning himself and sought a scientific solution. He commissioned cabinet maker Louis Soubrier to create a special elaborately upholstered "Love Chair" that allowed him to perform without his enormous girth crushing his paramours. That chair still exists, as do some replicas that can be purchased for around $70,000. The "Siège d'Amour" was equipped with a second cushion that supposedly allowed for sex with two women at once. It seems there was at least one type of chemistry in which the future King Edward VII was interested.

AN EXPERIMENT ON A BIRD

Will the bird live or die? That is the question we are left pondering when contemplating Joseph Wright's marvelous 1768 painting entitled *An Experiment on a Bird in the Air Pump*. The canvas on display in the National Gallery of London depicts a "natural philosopher," as scientists were called in those days, cranking a vacuum pump to remove the air from a glass bulb in which a cockatiel is struggling to breathe as onlookers express a range of emotions, from curiosity to horror. The demonstrator's hand hovers above a valve at the top of the glass globe.

Is he about to open the valve and allow the bird to survive, having made the point that air is necessary for life? Or is he ready to ensure that the valve is closed so that the experiment can come to its deadly conclusion?

The subject of the painting has a fascinating history that traces back to Evangelista Torricelli's classic 1643 experiment in which he filled a meter-long tube sealed at one end with mercury. With his finger over the open end he then inverted the tube and set it vertically into a basin of mercury. The column of mercury in the tube fell until it measured 76 centimeters in height. The space above the mercury now contained nothing, the first recorded case of a permanent vacuum! Torricelli's explanation was that we live in a "sea of air" that exerts a downward pressure the same way that water exerts pressure on a submerged object. The reason the mercury did not fall all the way down in the tube was because air was exerting a pressure on the pool of mercury in which the tube had been immersed. Torricelli, who was a pupil of Galileo, had invented the barometer, a device that measures air pressure.

Torricelli's experiment inspired Otto von Guericke, mayor of the German town Magdeburg, to create a device capable of producing a vacuum whenever desired. He managed to design the world's first vacuum pump, consisting of a piston in a cylinder equipped with one-way flap valves. A hand crank allowed the piston to move down and suck air out of a container to which the pump was attached. Think of the way a syringe can be used to produce suction.

To demonstrate his pump, von Guericke devised a pair of Magdeburg hemispheres that when fitted together formed a sphere about half a meter in diameter. Connecting a valve on one of the hemispheres to the pump allowed the air to be removed from inside the globe. Then came von Guericke's historic public display. In 1654, in front of a crowd that included Emperor Ferdinand III, two teams of fifteen horses each were attached to the sphere and were unable to pull the hemispheres apart until the valve was opened, allowing air to rush in.

Robert Boyle, with his view that matter was composed of elements that cannot be resolved into simpler substances, is regarded as one of the founders of modern chemistry. He was also a champion of arriving at conclusions based not on philosophy but experimentation. Boyle learned about von Guericke's vacuum pump and with the help of Robert Hooke constructed an improved version that allowed for experiments resulting in the formulation of Boyle's Law, stating that the volume of a gas is inversely proportional to its pressure, now drilled into the brains of high school students.

Boyle tested the effects of "rarified air" on various phenomena including sound, combustion, and magnetism. Then in a famous experiment, described in his 1660 book, Boyle placed mice, snails, flies, and birds in a glass globe that he then evacuated to demonstrate that air was necessary for life. He describes placing a lark in the chamber and watching as "the bird threw herself over two or three times, and died with her breast upward, her head downwards and her neck awry." This is precisely the experiment depicted in the Wright painting, replete with an accurate portrayal of Boyle's pump.

In the eighteenth century, science was emerging out of the darkness as the age of "enlightenment" unfolded and itinerant lecturers, more entertainers than scientists, performed experiments in front of paying audiences or in the homes of the wealthy. Wright's portrayal of the bird experiment affords a glimpse into the public's attitude towards science at the time and has relevance to today's controversies. Two girls are being urged, likely by their father, to pay attention to the experiment but appear to be reviled. An inquisitive boy looks on with curious anticipation of the outcome, a man with an obvious inclination towards science is timing the experiment, another is lost in thought, possibly contemplating the ethics involved. A young couple seem to have eyes only for each other, totally disinterested in the plight of the bird.

If we substitute "climate change," "endocrine disruptors," or "the COVID situation" for the bird, the painting can be seen to represent current views on scientific controversies. Some "onlookers" are focused on evidence, some prefer to ignore it, some are unsure of what is going on, and some live in ignorant bliss.

A boy in the painting holds a rope that seems to be swung over a beam to enable the empty birdcage to be raised or lowered. Is he moving it out of the way because a dead bird will no longer need it? Or is he lowering it to house the bird that will be allowed to live? The "philosopher" appears to be looking out of the picture, straight at us, as if imploring us to think about all the aspects of the situation. A wonderful, timeless painting. And if you look carefully, you will note a clever nod to von Guericke with a pair of Magdenburg hemispheres sitting in the shadows.

CAUSATION AND CORRELATION

Should you stop brushing your teeth? Statistics show that 98 percent of Canadians who develop COVID symptoms brushed their teeth two days before the onset of symptoms. Should you avoid saunas? Finland has one of the highest rates of heart disease in the world, and Finns own more saunas per capita than any other nation.

Actually, studies show that frequent saunas reduce the incidence of cardiovascular disease. So, what's with the Finns? They consume roughly 80 grams of fat a day, far more than the World Health Organization's recommended daily intake of 50 grams. They also love sugar, consuming close to double the 50 grams per day stipulated by WHO. Obviously, their high cardiovascular disease rate is more likely due to a poor diet than their love of saunas.

Although investigating such associations is at the very heart of science, the process can obviously be treacherous. The observation that the sun rises in the morning and sets at night led many to conclude that it circled Earth until Galileo and Copernicus came along. Since the 1950s, both obesity and levels of carbon dioxide in the air have risen significantly. Based on this association, one might therefore conclude that an increase in inhaled carbon dioxide causes obesity. However, science tells us that obesity is caused by an increase in calorie intake, not carbon dioxide inhalation.

There are many other similar examples. Does the shaking of tree branches cause wind, or does the wind cause trees to shake? If you live in the jungle, this may not be easy to answer. Wind and shaking trees always go together. Of course, sailors will know that there is wind in the middle of the ocean where there are no trees to cause it.

While associations cannot prove a cause-and-effect relationship, they can serve as a springboard for further investigation. The observation that lung cancer was seen more frequently in smokers spawned studies that proved smoking did indeed cause the disease. The observation that workers involved in the production of polyvinyl chloride (PVC) from vinyl chloride had an unusually high incidence of a rare type of liver cancer led to studies that clearly demonstrated the carcinogenicity of vinyl chloride.

In 1713, Italian physician Bernardo Ramazzini published the book *Diseases of Workers*, which is regarded as the first systematic investigation of occupational hazards. In a chapter entitled "Diseases of Cleaners and Privies," he described how these workers often suffered from a painful inflammation of the eyes and also noted reports of copper or silver coins in their pockets turning black. Ramazzini concluded that as the workers disturbed the excrement, some vapor that caused the eye irritation and also turned the coins black was released. A French commission set up to study the problem produced a report in 1885 that concluded, as Ramazzini had, that sewage emitted some sort of toxic gas, inhalation of which could even be lethal. Victor Hugo was apparently unaware of this danger since in his classic, *Les Misérables*, Jean Valjean trudges unaffected through the Paris sewage system as he carries the injured Marius to safety.

In 1772, Swedish apothecary Carl Wilhelm Scheele, who had developed a keen interest in chemistry, became the first person to isolate oxygen by heating mercuric oxide. Unfortunately for him, he didn't publish his work until 1777 in "Chemical Observations and Experiments on Air and Fire," by which time Joseph Priestley had published a paper describing essentially the same experiment performed in 1774. This earned Scheele the nickname "Hard Luck Scheele." Neither he nor Priestley recognized that the "air" they produced was an element. It was Antoine Lavoisier who correctly interpreted Priestley's experiment and is commonly recognized as the discoverer of oxygen. The credit really should be shared by the three men.

Scheele was also the first to produce hydrogen cyanide by heating potassium ferrocyanide with sulfuric acid and noted the almond-like odor of the gas. This time "Hard Luck Scheele" became "Lucky Scheele" because he escaped being poisoned by the cyanide. His luck persisted when he heated ferrous sulfide (fool's gold) with an acid and produced a gas he described as having a fetid smell. Scheele had made hydrogen sulfide, a gas that is more toxic than hydrogen cyanide, but its scent is so potent that Scheele likely left the premises and avoided poisoning. He did not identify the gas as hydrogen sulfide — that was determined by the French chemist Claude Louis Berthollet in 1776 — and it was Baron Guillaume Dupuytren who subsequently showed that this was the gas that caused the problems in the sewers of Paris.

How does this nasty gas form? Mostly by the reaction of sulfates with "sulfate-reducing bacteria" present in sewage. Sulfates in turn form when microbes act upon sulfur-containing proteins in decomposing organic matter, such as in sewage. As far as the coins are concerned, hydrogen sulfide reacts with silver or copper to form the corresponding sulfides that appear as a black deposit on the metal.

This phenomenon was also noted on copper air-conditioning coils in some Florida homes and in houses rebuilt after Hurricane Katrina. In both these cases domestic drywall had been in short supply and drywall was imported from China. Drywall is made of gypsum, or calcium sulfate, and it seems that the Chinese drywall was manufactured without proper preservatives, under humid conditions, and became contaminated with microbes that produced hydrogen sulfide from the sulfate. Many people complained of the smell of rotten eggs in their homes. Indeed, when eggs spoil, bacteria break down their protein components and release hydrogen sulfide. The amounts are way too small to cause poisoning, but they will remind you of a sewer.

One final observation. The vast majority of patients requiring intensive care in hospitals are unvaccinated. A spurious association? I think not. Oh, and keep brushing your teeth. That link with COVID is certainly bogus. Remember that correlation is not the same as causation.

CUCUMBERS AND PLASTICS

Trying to separate sense from nonsense for over four decades is quite an educational experience. I have learned a great deal, but perhaps the lesson at the top of the pile is that once you start scratching the surface of an issue, it invariably gets more complicated than it first seems. Rarely are issues black or white, they are various shades of gray. That is the case whether we are talking about electric vehicles, food additives, nutrition, cholesterol, medications, vaccination, climate change, insecticides, herbicides, personal care products, dietary supplements, space exploration, history, or plastics. Ahhh, plastics. They have become villains, the targets of emotional attacks by various bloggers who want plastics banned.

Let's get the nonsense out of the way. If you are going to ban plastics, you can forget about airplanes, cars, computers, cellphones, and you can close down hospitals. Obviously banning plastics is an absurd idea. However, given the deluge of plastic garbage and the frightening notion of microplastics building up in the oceans and possibly in our bodies, we have to engage in a risk-benefit analysis for specific applications. It is unreasonable to question the use of plastics in a heart-lung machine or in a surgical mask, but an English cucumber shrink-wrapped in plastic is a different story. Or is it?

The French government thinks it is different, and since January 1, 2022, cucumbers wrapped in plastic can no longer be sold. The plastic ban does not only apply to cucumbers, but to many other fruits and vegetables as well with some exceptions. Cut fruit and some delicate produce can still be sold in plastic for now. In June 2023, plastic packaging for fruits and vegetables was banned with the exception of berries, and by the end of 2024, such packaging must also be removed from asparagus, mushrooms, cherries, and some salads and herbs. Finally, by the end of 2026, all berries will have to be sold without plastic packaging. Sounds like a great idea, and it mostly is. But not in every case. Such as shrink-wrapped English cucumbers, in which case the benefits likely outweigh the risks.

Let's first look at the science of shrink-wrapping, a technology introduced in the 1960s. There are several different materials that can be used, the common feature being that they are all polymers. Thin films of polyvinyl chloride (PVC), polypropylene, or polyethylene can all be shrunk, with polyethylene being used the most widely. Picture a chain made of paper clips linked together. Each clip represents an ethylene molecule, and the chain is polyethylene. If you now drop the chain into your hand so that you can close your fist around it, it will coil more or less into a ball. A film of polyethylene is made of many such coils packed together. If this film is now stretched, the balls uncoil to form more or less straight chains that are maintained in this position by the small attractive forces that occur between atoms in adjacent chains. Nature, however, prefers randomness over orderliness, and if heat is now applied, the molecules become more vigorous and overcome the attraction between adjacent chains, which then proceed to coil again. The macroscopic effect here is that the film shrinks to fit snuggly around any object, be it a jar of tomato sauce, a medicine bottle, or a cucumber.

Why would a cucumber be shrink-wrapped? No producer has ever said, "We are not spending enough on our product, so let's increase our expenses by wrapping it in useless environmentally unfriendly plastic." The fact is that shrink-wrapping a cucumber can extend its shelf life by some 60 percent. There are several reasons for this. The wrap dramatically reduces moisture loss and prevents shriveling. It also reduces contact with oxygen in the air and therefore the rate of respiration. Fruits and vegetables continue to respire after being picked, meaning that their carbohydrate content reacts with oxygen to produce carbon dioxide and water, resulting in a change of texture. Oxidation is also responsible for other chemical changes that can affect nutrition. For example, shrink-wrapped broccoli loses far less of the glucosinolates thought to be responsible for the vegetable's health benefits than loose broccoli.

Produce that is damaged by moisture loss or oxidation ends up being discarded. It is estimated that about one-third of all food produced is wasted! The production and transportation involved in food that will never be used and the wasted agrochemicals leave a huge environmental footprint. Of course, plastic production also has an environmental impact, but it turns out that at least for cucumbers, the plastic wrap is responsible only for 1 percent of that impact. A cucumber that is discarded because of spoilage has the environmental impact of ninety-three plastic wraps. A cradle-to-grave analysis by the Swiss Federal Laboratories for Materials Science and Technology has concluded that unwrapped cucumbers have a five times greater negative impact on the environment than shrink-wrapped ones.

UNCLE FESTER

His appearance in the video is striking. Uncle Fester appears in a red devil costume replete with horns and a tail. This is not the Uncle Fester you may remember from the 1960s *Addams Family* sitcom, the one who sleeps on a bed of nails, feeds his plants with blood, and makes a light bulb glow by putting it in his mouth. This Uncle Fester is a clandestine chemist, real name Steve Preisler, who acquired the nickname during his undergraduate days as a chemistry major because of his penchant for doing crazy things in the lab. After graduating, he put his chemical knowledge to use by converting ephedrine, at the time readily available in various cold remedies, to methamphetamine, "speed" or "crank" in street language. For this he got pinched and sentenced in 1984 to five years in prison.

In jail he happened to watch a television exposé of "terrorist publishers" who released books with instructions for making explosives. An idea was born! He would get back at the authorities, who in his mind had imposed an unfair sentence for making just a few grams of meth. He would train an army of chefs in the art of cooking crank! Preisler borrowed a typewriter and under the pseudonym "Uncle Fester" cranked out his cult classic, *Secrets of Methamphetamine Manufacture*, in which he detailed methods of synthesizing meth and its required precursors from simple, mostly readily available chemicals. The book, speckled with whimsical anecdotes, gave specific instructions easily followed by anyone with a basic background in organic chemistry. Preisler called his work "good clean chemist fun" but authorities labeled him as "the most dangerous man in America."

Fester is still at it, as Vice Media's 2021 video by him makes clear. He seems to relish his infamy as a disseminator of "forbidden knowledge" as evidenced by the devil costume, but there are several disturbing issues here. While the procedures as shown in the video are way too haphazard to be followed, it does repeatedly refer to his *Secrets of Methamphetamine Manufacture*, now into the eighth edition. The book is even available on Amazon for would-be chemists trying to follow in Fester's footsteps. But there is something else. His reckless spilling of unmeasured amounts of chemicals, smoking while dealing with flammable substances, lack of ventilation, callous use of toxic mercury compounds, and disposal of chemicals in the toilet all demonstrate a total disregard for safety. This is just not the way to do chemistry of any kind! A real smudge on the face of a most useful science.

When asked why "America's favorite clandestine chemist" is not afraid the video will trigger his arrest, Preisler claims that he has deliberately left out the last step in the synthesis, ensuring that nobody will be able to make meth based on what they have seen. He also points out that murders are routinely shown on television and wonders why anyone would object to a little chemistry being shown. Well, the chemistry being shown is not innocuous and can lead to real deaths. His argument that large-scale meth manufacture today is in the hands of Mexican cartels and authorities should not be bothered by little guys cooking up a few grams for their own occasional use is a vacuous one. Preisler attempts to justify this by referring to the classic dictum that "only the dose makes the poison," although curiously, as a chemist, he is unaware that this was first voiced by Paracelsus. Evidence, however, indicates that there is plenty of meth being synthesized in clandestine labs north of the border that supply addicts, not "occasional" users.

It is little wonder that creators of the television series *Breaking Bad* sought out "Uncle Fester" for advice. That popular show depicted a high school chemistry teacher who is diagnosed with cancer and worries about his family's financial situation after his demise, which seems imminent. He decides to descend into the criminal underworld and cook some meth to make money. Preisler says he helped with the scripts and design of the equipment used. He even suggests that the character of Walter White was based on him! That is possible because by the time the series aired in 2008, "Uncle Fester" and his books had garnered a large following. He had written about synthesizing various other controversial chemicals, including LSD, nitroglycerin, and phosgene. In *Silent Death* he describes the synthesis of the nerve gas sarin. That book was found in the belongings of the terrorist gang that unleashed an attack with this gas in the Tokyo subway system in 1995, killing thirteen people. Preisler's bizarre comment was "that was regrettable, but it is nice to know the recipe works."

Today, Preisler mocks and taunts the authorities he claims are out to get him even though he is not himself producing anything. He also vigorously attacks "copyright piracy," claiming stolen copies of his works can be illegally downloaded. He pontificates about the evils of stealing other people's work! What about taking chemical reactions that have been developed by researchers for the advancement of chemistry and hijacking them to teach covert chemists how to make drugs that cause misery and death?

The Addamses' Uncle Fester never did anything that nefarious. He didn't even use drugs. When he had a headache, he just placed his head inside a large screw press and proceeded to tighten it. Sometimes he even used the screw press simply for enjoyment. Quite different from the clandestine chemist Uncle Fester, who enjoys screwing with people's lives. He is a criminal and a blight on the face of chemistry.

INFLAMMATION INFORMATION

Ouch! You just stubbed your toe. It quickly swells, gets hot, turns red, and it hurts. You may curse, but you are actually experiencing the beginning of the healing process. Inflammation is the body's attempt to right the wrongs caused by physical injury, infection, or exposure to toxins. Thanks to acute inflammation, you will soon be back to kicking without pain. Long-term or chronic inflammation, however, is a different story. That may have you kicking the bucket.

Way back in the first century AD, Roman encyclopedist Aulus Cornelius Celsus produced a comprehensive medical work, *De Medicina*, in which he described the use of opiates to counter pain, explained that fever was the body's attempt to restore health, and introduced the tetrad of rubor (redness), calor (heat), tumor (swelling), and dolor (pain) as the cardinal signs of a condition we now refer to as inflammation. Of course, knowledge of physiology at the time was too rudimentary to offer an explanation for what was going on, but it was clear that inflammation was a prelude to healing.

Today, we know that redness is caused by dilation of blood vessels in the area of injury as a result of an increased blood flow that can also be sensed as heat since blood is warm. The blood delivers white blood cells (neutrophils) to clean up the cellular debris caused by injury, antibodies to destroy bacteria and viruses, and clotting factors that prevent the spread of infectious agents through the body. Chemical mediators of inflammation, such as histamine and cytokines, change the permeability of blood vessel walls to allow white blood cells to diffuse from the bloodstream into injured tissues. Prostaglandins rush to the scene to elevate temperature and impair microbial activity. As fluid carrying white blood cells enters injured tissue from the bloodstream, it causes swelling, which in turn causes pain.

Finally, after the white blood cells have managed to gobble up the remains of injured tissues, and antibodies have neutralized microbes, healthy cells begin to multiply. Pain resolves, swelling subsides, and memory of the acute inflammation fades.

Now for a more worrisome scenario. Inflammation is an essential response for dealing with various forms of assault on the body, but it is not always perfectly controlled. Cholesterol deposits in arteries, foreign substances such as silica dust, as well as some infectious organisms may resist the body's attempts to eliminate them and precipitate a continuous attack by white blood cells. The immune system can also make a mistake and launch an inflammatory strike against a normal body component resulting in an autoimmune disease such as rheumatoid arthritis, multiple sclerosis, celiac disease, or type 1 diabetes. Even some foods or specific components can be seen as an enemy to be neutralized. The result is a chronic, low-grade inflammation that is associated with cardiovascular disease, diabetes, and some cancers.

Obviously, chronic inflammation is undesirable, but how do we know when it is present and what can we do about it? The inflammatory activity of white blood cells is associated with the release of chemicals into the bloodstream that can serve as markers of inflammation, with the major ones being interleukin 6 (IL-6), high-sensitivity C-reactive protein (hs-CRP), fibrinogen, homocysteine, and tumor necrosis factor alpha (TNF-alpha). These markers rise with obesity, smoking, inactivity, sleep deprivation, and poor diet.

The diet connection has received much attention because it is a readily modifiable lifestyle factor. Based on an extensive literature search of cell culture studies, animal experiments with specific nutrients, and human epidemiological studies in which the relationship between inflammation markers and diet was determined, researchers have developed a dietary inflammatory index (DII). Through a complex formula, various foods and forty-five specific nutrients are assigned numerical values based on how they affect inflammatory markers. Sugar, trans fats, refined carbohydrates, omega-6 fats, red and processed meat are classified as inflammatory, while fiber, vitamin E, vitamin C, beta-carotene, magnesium, and moderate alcohol intake are anti-inflammatory. A food frequency questionnaire can then be used to calculate the anti-inflammatory effect of a specific diet.

It comes as no great surprise that the typical Western diet with its high red meat, full-fat dairy, refined grain, and low fruit and vegetable consumption is associated with higher levels of CRP, IL-6, and fibrinogen. By contrast, the Mediterranean diet that features whole grains, fruits, vegetables, fish, olive oil, moderate alcohol consumption, and little butter or red meat is associated with lower levels of inflammation.

When DII scores were calculated in a study of some 5,000 adults, those who ranked in the top quarter, meaning they consumed the most inflammatory foods, had much higher CRP levels than those in the bottom quartile. This indicates that a DII score can indeed predict whether a specific diet is linked with inflammation. Even more significantly, meta-analyses, essentially the pooling of relevant studies, found an association between a low DII score and protection against cancer as well as cardiovascular disease.

Does this mean that we should all be testing for inflammatory markers in our blood to know whether we are at risk of chronic low-grade inflammation? Not unless a physician suspects, based on symptoms, that there may be some disease process. Otherwise, what you would do in response to elevated markers is what we should all be doing anyway. Exercise, watch our weight, minimize highly processed foods, and emphasize whole grains, fruits, vegetables, beans, lentils, nuts, fish, and olive oil. As far as the plethora of dietary supplements flooding the market with claims of "reducing inflammation" go, the only documented reduction will be to your bank account.

WINE AND HEALTH

"When I left France, I couldn't imagine that there was a population in the world that didn't drink wine with meals." That remark by a young Serge Renaud, who had come to study veterinary science at the University of Montreal in 1951, was prompted by his observation of dietary habits in Quebec. He was also surprised by the low consumption of fruits and vegetables and the high intake of saturated fats. Renaud also noted statistics about the high rate of heart attacks in Quebec and was particularly taken by reports of young hockey players developing coronary disease. Was there a causal link between diet and heart disease? Exploration of that possibility would come to dominate Dr. Renaud's career and lead to him being anointed as the father of the "French Paradox."

After graduating first in his class from veterinary school, Renaud enrolled in a PhD program in experimental medicine. His research focused on thrombosis, the blockage of arteries or veins by blood clots, a problem he would pursue as head of the laboratory of experimental pathology at the Montreal Heart Institute. Dr. Renaud concluded that the tendency of tiny cells called platelets to aggregate and form clots was the main cause of heart attacks and suspected that this was related to diet.

In 1973 he returned to France, and at the National Institute of Health and Medical Research began to study the behavior of platelets and observed that aggregation was reduced in rats that had been fed alcohol. This was intriguing, especially because he had heard that the famous Framingham study in the U.S. had found that alcohol protects against heart disease. That information had not been publicized because the National Institutes of Health was concerned that if people learned that a little alcohol was protective, they would conclude that more is better.

To explore the diet connection further, Dr. Renaud organized a mobile laboratory to travel around France to take blood samples and determine platelet reactivity. The goal was to investigate a possible link with heart disease given that a geographic variation had been observed in France, with a high incidence of cardiovascular conditions in the Moselle region and a low rate in Provence. Indeed, as it turned out, the disease incidence was connected to platelet aggregation. But why did platelet reactivity vary regionally? Dietary questionnaires people had filled out suggested that saturated fats and smoking were likely to cause blood clots, while calcium, polyphenols, wine, and alpha-linolenic acid in the diet were protective.

That was interesting, especially since the famous Seven Countries Study by Ancel Keys had also investigated a possible connection between diet and heart disease and had found that the inhabitants of the island of Crete had the lowest rate despite having a relatively high level of blood cholesterol. Something was protecting the Cretans! Could it be the alpha-linolenic acid found in the snails, nuts, and purslane that are prominently featured in the Cretan diet?

To investigate this possibility Dr. Renaud launched the Lyon Diet Heart Study in which men who had suffered a heart attack were divided into two groups. The control group followed the usually recommended low-fat diet, while the experimental group ate a diet modeled on that in Crete. No butter, cream, or milk and very little meat. Lots of vegetables, fruit, bread and whole grains. And wine with meals. They also ate a specially formulated margarine rich in alpha-linolenic acid since walnuts, snails, and purslane were not easily incorporated into the diet. The trial was stopped early because a striking difference between the groups was noted just two months after diet modification began. Compared with the "prudent" diet, the "Cretan" diet reduced cardiac death by 76 percent, non-fatal heart attacks by 73 percent, and total death by 70 percent.

By 1991 Dr. Renaud's research had garnered publicity and came to the attention of the CBS program *60 Minutes*. In a segment entitled "The French Paradox," he was asked how it was that the rate of heart disease in U.S. was more than three times greater than in France despite the French love of fatty cheeses, cholesterol-laden goose liver, and buttery croissants? Renaud suggested that there were several differences. The French didn't eat between meals, had the main meal at lunch, ate lots of cheese, and of course drank wine. He explained that his studies of rat poop had shown that calcium in cheese can tie up fats and cause them to be excreted, and he also described his lab studies of wine decreasing platelet aggregation. Dr. Renaud was careful to point out that while he believed a moderate amount of wine to be protective, certainly more was not better. The next day wine sales in the U.S. soared!

Since that classic *60 Minutes* segment a large number of studies have examined the link between wine consumption and health. You can pick and choose among these studies to show either that that there is no safe amount of alcohol or that moderate consumption protects against heart disease. The only point on which there is universal agreement is that more than "moderate consumption," usually taken to mean more than one to two servings of alcohol a day, is detrimental to health. Then there is the niggling issue of alcohol being carcinogenic. Indeed, the International Agency for Research on Cancer (IARC) has placed alcohol in Group 1, reserved for substances known to cause cancer in humans. In spite of all the studies, whether the phrase "to your health" while clicking wineglasses is or is not backed by science remains an open question.

PROBLEMS WITH PALM OIL

It was a welcome change for the welfare of coronary arteries. It was not so good for the welfare of orangutans. That change was the substitution of palm oil for trans fats in numerous consumer products. Back in the 1960s researchers began to associate the consumption of saturated fats, as found in butter, shortening, and meat, with deposits in the coronary arteries. Such deposits were in turn associated with heart disease. Industry reacted by looking at vegetable oils, low in saturated fats, as possible replacements for butter and shortening. The problem was that most vegetable oils are liquids. However, there was an apparent solution. If the unsaturated fats in a vegetable oil were subjected to treatment with hydrogen gas, the carbon-carbon double bonds that characterize such oils would take up the hydrogen and be converted to single bonds as found in saturated fats. There would be no point in eliminating all the double bonds, since that would result in a fully saturated fat, just like butter. But "partial hydrogenation" resulted in a product that was solid but less saturated than butter. Margarine began to replace butter in shopping carts, and labels on baked goods hyped that they were made with vegetable shortening.

By the early 1990s, there were suggestions that we may have gone from the frying pan into the fire. "Partial hydrogenation" also had an effect distinct from converting double bonds into single ones. Normally the carbon atoms attached to those involved in the double bond are on the same side of the double bond in what is referred to as a "cis" configuration. Hydrogenation can alter the geometry of the double bond so that the attached carbon atoms end up on opposite sides of the bond, resulting in "trans" fats. These turned out to be even more capable of causing deposits in coronary arteries than saturated fats! The race to eliminate the dastardly trans fats was on.

Most vegetable oils are liquids at room temperature, but coconut oil, palm oil, and palm kernel oil are exceptions due to their high saturated fat content. Since their saturated fat content is still lower than butter's, manufacturers turned to these fats to replace partially hydrogenated fats in processed foods. Palm oil was particularly attractive because it resists degradation when used for frying and has a consistency that is ideal for producing baked goods. And crucially, it is cheap to produce since the oil palm tree that bears the fruit from which the oil is produced has the highest yield per acre of any oilseed crop. Furthermore, the oil extracted from the kernel of the fruit is high in lauric acid, which can be converted into lauryl alcohol, the key compound for the production of sodium lauryl sulfate, the most widely used surfactant in the world. Surfactants are key ingredients in detergents, forming a link between water and greasy deposits. Sodium lauryl sulfate has the added benefit of dislodging soil from surfaces through its ability to boost foam. More recently, palm oil has been used to produce biodiesel fuel, further increasing demand.

That demand has consequences. Producing more palm oil requires planting more oil palm trees, which requires more land. Unfortunately, in Malaysia and Indonesia, the prime producers, the quest for more planting acreage leads to deforestation, often by setting fire to rainforests. The carbon dioxide released by these fires is a significant contributor to the greenhouse effect, which of course is a concern. But there is another issue. Expanding palm tree plantations, particularly in Borneo and Sumatra, are encroaching on the habitat of wildlife. These are the only two locations where orangutans are still found, and the species is now critically endangered. The same goes for the Sumatran tiger and rhinoceros. Many other rainforest animals may face the same fate as their habitat is replaced by palm trees. Turning away from the use of palm oil is not viable since it is an ingredient in hundreds and hundreds of consumer products ranging from cookies and cereals to lipstick. However, the oil can be produced in a sustainable fashion on plantations that have not relied on deforestation.

Nutritionists have also raised alarm about consuming foods containing palm oil on account of its saturated fat content, although in recent years the relationship between heart disease and saturated fats has blurred somewhat and these fats may not be quite as villainous as it once seemed. As a general rule, though, palm oil is often found in processed foods, a class that has been linked with ill health and premature mortality.

Now producers of the most widely used vegetable oil in the world may have something else to worry about thanks to some recent headlines: "Palm oil used in chocolate spreads, cooking oil may fuel cancer spread." "Palm oil linked to increased cancer risk, study finds." "Acid from palm oil linked to cancer spread." Scary stuff, especially considering that the research to which the headlines refer was published in *Nature*, one of the world's most respected science publications. However, the headline writers missed a key feature of the study. It was done in mice! Researchers inoculated mice with human squamous carcinoma cells and then fed them a diet high in palm oil or olive oil. Once tumors developed, cells were extracted and transplanted into mice that were fed a standard lab diet. The tumors in this second group of mice that originated in the palm oil–fed mice metastasized much more quickly than the tumors from the olive oil–fed mice.

While this is valuable research in trying to unravel the mysteries of cancer metastasis, it has no direct implication for humans. First, people are not giant mice. Also, the amount of palm oil the mice were fed was far more than that found in the human diet. And there was no demonstration that palm oil had an effect on the growth of primary tumors, only on the ability of the tumors to spread once transplanted to other mice. The paper's senior author did not claim that the research had direct meaning for people, but opined that patients with metastatic cancer might benefit from a diet poor in palmitic acid. And that might benefit the orangutans as well.

ISSUES WITH THE PEOPLE'S CHEMIST

Ill-informed, self-declared experts yelping from atop their soap boxes are a dime a dozen these days. But it is one thing to hear nonsensical rhetoric about the need to avoid substances with unpronounceable names from some scientifically illiterate blogger, and quite another to witness a legitimate chemist urging people to adopt a "chemical-free" lifestyle. Yet those are the exact words used by Shane Ellison who has anointed himself "The People's Chemist." Needless to say, "chemical-free" is a dually absurd expression. For one, nothing, save a vacuum, is chemical free since chemicals are just the building blocks of all matter. Secondly, the message implies that "chemicals" are synonymous with "toxins" or "poisons." Nonsense! Chemicals are not good or bad, their specific use has to be judged on their merits as determined through proper scientific investigation. How the word "chemical-free" can come out of the mouth of someone with a master's degree in organic chemistry is a true mystery.

While "chemical-free" can be construed to be benign gibberish, urging people to "Ditch Your Meds" is anything but. Ellison, who once worked for a pharmaceutical company, now tells people that the time has come to give up their blood pressure, thyroid, cholesterol-lowering, and antidepressant meds because these are evil concoctions foisted on the world by a heartless profit-driven medical industrial complex. Incredibly, casting insulin aside is also included in this deplorable, foolhardy advice.

Ellison's vengeful attacks also target COVID vaccines, which according to him contain a "chemical s***storm to fight the common cold dressed up as COVID-19." He even links the vaccines to the production of Zyklon B, the notorious cyanide-releasing chemical used in the Nazi gas chambers. Zyklon B was made by the IG Farben Company, which Ellison says "after the war broke off into Bayer and Merck which became Moderna." That is ludicrous and even the history is wrong. Merck was around long before IG Farben, and Moderna is not a Merck spin-off. Just another example of the lengths to which rogues will go to gain a following. Not surprisingly, Ellison also claims that the "HIV/AIDS hypothesis was one hell of a mistake" and that "Dr. Fauci and other Pharma-fueled scientists" made millions from selling treatments that were unnecessary. Just totally mindless blather.

By now you have probably guessed that this sage offers an alternative to the drugs that he says should be ditched. A bevy of "chemical-free" supplements is available for purchase through his website. For example, after dissing vaccines and claiming they are not needed for a virus that doesn't exist, he pushes his own Immune FX, which contains an extract of two plants, *Andrographis paniculata* and coriander. Apparently, plants do not contain chemicals. A literature search for the medicinal value of these plants comes up with some evidence of coriander having antibiotic effects in the gut of broiler chicks, and some andrographis compounds having anti-inflammatory and antioxidant activity when tested in cell cultures. There is zero evidence of Immune FX having been put to a test in human trials.

A supplement branded as Serotonin FX suggests that it boosts levels of serotonin, a neurotransmitter that can indeed have an antidepressant effect as clearly demonstrated by the use of prescription serotonin reuptake inhibitors (SSRIs). Ellison's formula contains the amino acid L-tryptophan that, once again, is not considered to be a chemical. While L-tryptophan is indeed the body's precursor to serotonin, there is no clinical evidence of equivalence to SSRIs. Patients who have been prescribed these drugs are wading into deep waters should they replace them with Serotonin FX.

The other supplements hyped by the People's Chemist are similarly devoid of evidence. His answer to pain is Relief FX, with ingredients extracted from white willow bark and ginger root. The willow bark contains salicylic acid that has pain-relieving properties but also irritates the stomach. That is why it has been replaced by acetylsalicylic acid, better known as aspirin. Synthetic aspirin is superior to natural salicylic acid, so going back to willow bark makes no sense. Ginger is a chemically complex mixture of dozens of compounds, some of which do have analgesic effects, but the label on Relief FX yields no information about what ginger-derived compounds are present or in what dosage.

People who are concerned about testosterone levels are offered Raw-T, an extract of sarsaparilla root. But they will be disappointed if they replace their prescription testosterone with this concoction. Testosterone is a steroid, and sarsaparilla does contain related compounds called sterols; however, these are not converted in the body to testosterone. Neither is there any evidence that Preworkout, advertised as "Faster Higher Stronger in 59 Minutes Without Chemicals," delivers the goods. Never mind that citrulline, tyrosine, Hawthorn extract, yerba mate, and huperzine A are absurdly described as not being chemicals, the claim that any substance can make someone faster, higher, or stronger in fifty-nine minutes sticks in the craw.

The People's Chemist curiously uses a microscope as his logo, an instrument not commonly used by chemists. He also describes how he walked away from an "award-winning career as a medicinal chemist" when he discovered that people taking cholesterol meds, cancer drugs, and blood thinners were just victims of pharma's insidious marketing practices. Indeed, there are some reprehensible practices, and they should be brought to light, but that does not mean that all drugs should be tarred and feathered with blatantly farcical arguments. Ellison says that he poured blood, sweat, tears, and years into becoming a chemist. He could also have used a bit of sense poured into his head. With an attempt at humor, he adds that, actually, there were no tears because chemists don't cry. Wrong! His prattle about a "chemical-free" lifestyle can bring a tear to any chemist's eye. The only question is whether it is from laughing or crying.

BATS, VAMPIRES, AND LONGEVITY

Charles Darwin was prone to seasickness. That is why the *Beagle* made intermittent stops as the ship cruised along the Chilean coast in 1835.

It was during one of these stops that Darwin famously encountered a "vampire bat." Europeans at the time had already heard about the existence of such creatures from early explorers of the Americas, and in 1790, zoologist George Shaw had coined the term "vampyre bat." However, before Darwin, there had been no first-hand account of the bat dining on a blood meal. The naturalist described how with its extremely sharp teeth, *Desmodus rotundus* (the bat's scientific name) taps into an animal and uses its tongue to lap up the blood that trickles out. No sucking is involved!

Shaw's use of "vampyre" to describe the bat was clearly based on the mythology that had been spreading from Eastern Europe about the dead rising from the grave and nourishing themselves on the blood of the living. That myth had become so entrenched that in 1751 it prompted an investigation by Antoine Calmet, a Benedictine scholar who published his findings in a "Treatise on the apparitions of Spirits and on Vampires or Revenants of Hungary, Moravia, Bohemia and Silesia." He described reports of "men who have been dead for several months, come back to earth, talk, walk, infest villages, ill use both men and beasts, suck the blood of their near relations, make them ill, and finally cause their death." The only way to stop the hauntings was "by exhuming the bodies, impaling them, cutting off their heads, tearing out the heart or burning them." Calmet was taken by the number of accounts of vampirism, and while he attempted to refute them with suggestions that the "undead" were actually victims of malnourishment or disease, he wasn't very convincing. Indeed, the myth has still not had a stake driven through its heart in modern-day Transylvania.

The romanticized fictional version of the vampire appeared in 1819 with John William Polidori's *The Vampyre*, followed by Rymer and Prest's *Varney the Vampire*, in which the bloodthirsty villain is depicted for the first time as having bat-like wings. Of course, it was Bram Stoker's *Dracula*, published in 1897, that became the prototype for subsequent versions of the sanguinary count and introduced the story lines of immortality and transformation into a bat.

The popularity of the book and the various movie versions of Dracula have unfortunately saddled bats, especially vampires, with a negative image. Vampire bats actually make up only a tiny segment of the nearly 1,000 bat species and are limited to Latin America. Nobody in Europe or North America need worry about being bitten! In any case, vampires do not make a habit of attacking humans, although someone sleeping barefoot outdoors on a very dark night might get nipped on the toe, not on the neck. Domestic animals such as chickens, cows, horses, and pigs are the usual prey, which presents a problem because vampire bats can cut into ranchers' profits by infecting livestock with rabies. This has resulted in misguided vampire control programs using poisoned bait such as bananas. While all other bat species do eat fruit, vampires consume only blood. Innocent bats thus become victims, resulting in ecological consequences given that they eat insects and serve as pollinators. They are the prime pollinators of the agave plant. No bats, no tequila!

Bats live in colonies, often numbering many thousands. And they produce copious amounts of poop! Extremely useful poop! The high content of nitrogen, phosphorus, and potassium makes bat guano an ideal fertilizer, which explains why before the discovery of oil in Texas, it was the state's most important export. Potassium nitrate, better known as saltpeter, can also be extracted from guano. Along with sulfur and charcoal, it is a critical component of gunpowder and explains why during the American Civil War the protection of bat caves was of paramount importance to the Confederate army. Northern ships had blockaded southern ports and prevented the importing of gunpowder, making bat guano the only source of saltpeter.

Bat feces are flammable and also yield methane gas on decomposition. A group of settlers from Kentucky who came to Texas to mine guano from a bat cave in 1854 found out about this the hard way. The story is that lightning struck near the mouth of the cave and ignited the methane that had built up. There was a reverberating explosion. The name of the town they founded? Blowout, Texas.

The real blowout about bats, though, is their longevity. Not quite immortal like Dracula, but they live at least four times longer than other similarly sized mammals. Longevity in mammals often correlates with size, with an elephant living about seventy years and a mouse, about the size of a bat, just a couple. Bats on the other hand can live for decades. A bat in Siberia that had been tagged with a microchip set the record at forty-one years! So, can bats hold the secret to mammal longevity? Adjusting for size, if humans lived as long as bats, we could look forward to a life expectancy of some 240 years!

Researchers are obviously interested in exploring the surprising longevity of bats and are examining various possibilities. It seems that in bats, the telomeres, those protective segments of DNA at the tip of chromosomes, do not shorten with age as in other mammals. This offers better protection against chromosomal damage when cells divide. Hibernation may also play a role, and who knows, maybe even sleeping upside down. Finally, bats may even supply us with novel drugs. An enzyme in their saliva that prevents their victim's blood from coagulating, making for easier slurping, is being studied as a possible treatment for strokes caused by blood clots. It has been imaginatively named Draculin.

HITLER AND PROBIOTICS

Heinrich Hoffmann had contracted gonorrhea and sought help from Dr. Theodor Morell, who had become a fashionable physician in Berlin, treating the rich and famous with vitamins, herbs, and various animal gland extracts. Apparently Hoffmann was satisfied with what-ever treatment he received because he recommended Morell to his boss, who happened to be Adolf Hitler. Hoffmann was the Führer's official photographer and trusted member of his inner circle.

Hitler suffered from chronic intestinal problems and was increas-ingly irritated by the lack of help from his physicians and agreed to meet Morell, who treated him with Mutaflor, a preparation that contained live bacteria. The Nazi leader was so elated by its effects that he made Morell his personal physician much to the dismay of many in his entourage. Hermann Göring, who founded the Gestapo, and Heinrich Himmler, the main architect of the Holocaust, dismissed Morell as a quack and an opportunist.

Indeed, Morell had a less than stellar history, having falsely claimed to have studied under Ilya Mechnikov, who was awarded the 1908 Nobel Prize in Medicine and Physiology for his work on immunity. Morrel had concocted the connection with Mechnikov to enhance his own reputation. The Ukrainian scientist had garnered fame thanks to his discovery of a type of immune cell that has the ability to engulf and ingest harmful substances such as bacteria and dead or dying cells. These cells would be termed "phagocytes," from the Greek "phagein" meaning "to eat," and "cyte" denoting "cell."

In this context, Mechnikov had studied microbes in the gut and theorized that increasing the population of harmless bacteria might curb the growth of disease-causing organisms. Since all bacteria in the gut compete for the same food supply, the harmful ones would be starved and end up being gobbled up by phagocytes. Mechnikov believed that lactic-acid-producing bacteria, such as the ones found in sour milk, were the key to good health, and even attributed the supposed longevity of Bulgarian peasants to their yogurt consumption. He himself drank sour milk every day, thereby laying the foundation for the use of what today we call "probiotics," from the Latin "for life." By definition, these are "live microorganisms that when administered in adequate amounts confer a health benefit on the host."

Professor Alfred Nissle was an infectious disease specialist who was familiar with the work of Mechnikov. In 1917 he managed to isolate a strain of *Escherichia coli* bacteria from the feces of a soldier who was resistant to an epidemic of diarrhea, likely caused by *Shigella* bacteria. He surmised that this strain, which would be known as "*E. coli* Nissle 1917," was a "probiotic" that had crowded out *Shigella*. Nissle introduced this strain into practice as Mutaflor, claiming that it would treat intestinal diseases, which it actually did with varying degrees of efficacy. This is what Morrel administered to Hitler, apparently with success.

Long before Mutaflor, people had been consuming probiotics through fermented foods such as sauerkraut, miso, tempeh, kefir, kimchi and yogurt. These are produced through the controlled growth of microbes such as yeasts and bacteria and have historically been deemed to be healthy due to their content of live bacteria. In the 1800s, patients with all sorts of complaints flocked to Dr. John Harvey Kellogg's "Sanitarium" in Battle Creek, Michigan, where they would be treated with "lactic ferment" — in other words, yogurt. Kellogg was an early disciple of Mechnikov and extolled the virtues of lactobacilli as a means of "driving out poison-forming germs." He advocated the administration of yogurt both orally and by means of enemas, "thus planting the protective germs where they are most needed and may render most effective service."

Kellogg and Mechnikov, it seems, were on the right track. Since the 1990s, research into probiotics has exploded with the composition of bacteria in the gut, the so-called microbiota, being linked with type 2 diabetes, hypertension, depression, Alzheimer's disease, colorectal cancer, obesity, and inflammatory bowel disease such as colitis, which medical historians believe was the cause of Hitler's gut problems. It is therefore conceivable that he could have been helped by Mutaflor. As a consequence, he came to rely on all the other medications that Morell suggested for his various other complaints that included headaches, constant colds, and insomnia.

Morrel kept meticulous records of his "Patient A." These survived the war and have been thoroughly analyzed. Between 1941 and 1945, he treated Hitler with twenty-nine different injections and sixty-three kinds of oral tablets that included codeine, cocaine, testosterone, sulfonamide, oxycodone, strychnine, belladonna, bile extracts, morphine, and barbiturates. Vitamultin was Morrel's special concoction of vitamins produced in one of the highly profitable drug firms he owned. When

Hitler was injected with this, he claimed to feel refreshed and invigorated. Himmler, who was suspicious of Morrel, secretly had one of his Vitamultin pills analyzed and learned it contained methamphetamine!

Many of the drugs administered by Morrel have psychoactive properties and possibly contributed to Hitler's increasing paranoia, anxiety, and perhaps even to some poor military decisions. One wonders how history would have been altered had Hitler had not had success with Mutaflor, a preparation still available today in Germany for treatment of ulcerative colitis, chronic constipation, and diarrhea in newborns. It was the relief afforded by Mutaflor that led to the Führer's utter trust in Dr. Morrel and the medications he meted out, many of which were questionable and untested. However, any suggestion that the nightmare Hitler inflicted on the world was somehow linked to drugs he was taking should be dismissed. Hitler was simply evil personified.

MOLECULES AND MIRRORS

Everything has a mirror image save perhaps a vampire. Place a spoon in front of a mirror and imagine that somehow you were able to reach into the mirror and pick up its mirror image. If you now compare the two items, they would be identical. Now imagine taking a glove off your left hand and placing that in front of a mirror. If you were able to pick up the mirror image, you would find that it fits the right hand, not the left. If you superimposed one on the other, the thumbs would point in opposite directions. What we have are non-superimposable mirror images! What is the criterion for this property? A lack of symmetry! Any item that is symmetrical will be identical to its mirror image, while items that lack symmetry will have a non-superimposable mirror image. Some items, like pairs of gloves or shoes, actually exist as non-superimposable mirror images. So do some molecules.

The commercial synthesis of the popular pain reliever ibuprofen (Advil, Motrin) yields two non-superimposable mirror image forms, commonly referred to as "enantiomers." It turns out that only one of the two has biological activity, not an unusual phenomenon for reactions that occur in the body. Often molecules have to engage with enzymes or fit into receptors on cells, both of which require interaction with proteins that are "chiral" or have "handedness." One enantiomer will fit, like a left glove fitting the left hand. The other will be like trying to put a left glove on the right hand. With ibuprofen, the inactive version is harmless, so the medication is sold as a mixture of the two enantiomers, known as a racemic mixture.

In some instances, such as with Dopa, the drug that is widely used to treat Parkinson's disease, only one enantiomer is converted in the body to dopamine, the molecule that is in short supply in the disease. The other enantiomer is not only useless but gives rise to serious side effects. In this case, it is important to be able to produce only the active form, known as L-Dopa, which can be done by so-called asymmetric synthesis. William Knowles shared the 2001 Nobel Prize in Chemistry for developing this process.

Sometimes marketing rather than efficacy drives the introduction of a single enantiomer drug. The anti-ulcer medication omeprazole (Prilosec, Losec) made a fortune for its manufacturer AstraZeneca, but it was set to go off patent in 2001. How to beat off competition from generics that were expected to flood the market? AstraZeneca came up with a plan. Omeprazole is a chiral drug, but in this case, both enantiomers are active so historically it was marketed as a racemic mixture. If one enantiomer could be shown to be somehow preferable, it could be marketed as a new drug and receive patent protection. Some clever data mining revealed that one enantiomer, christened esomeprazole, was broken down a touch more slowly in the body. This was enough to obtain a patent and esomeprazole went on the market as Nexium. This meant that half as much Nexium would have the same efficacy as the usual dose of omeprazole, which was 40 milligrams. But AstraZeneca recommended Nexium also at 40 milligrams, so people naturally did feel more relief. Critics argued that the same result can be had at a fraction of the cost from just doubling the dose of racemic omeprazole. Nexium turned out to be a big win for AstraZeneca but not necessarily for the public.

Now there is an interesting story emerging with another chiral drug, ketamine, a widely used anesthetic and painkiller that also has been found to have an antidepressant effect. Although there has been no formal approval for its use as an antidepressant, physicians have been prescribing it off-label to patients who were not helped with the standard medications such as the selective serotonin reuptake inhibitors (SSRIs). Since ketamine is an older drug and is off patent, there has been no motivation for pharmaceutical companies to fund studies about its antidepressant effect. At least not until 2020, when Janssen Pharmaceuticals decided to explore the possibility that one of the ketamine enantiomers may have superior efficacy. The single enantiomer could then be marketed as a novel drug and receive patent protection.

The first challenge was to either develop a synthetic process yielding only the single enantiomer or to find a means of separating the components of the racemic mixture. The latter turned out to be more practical. A technique known as chiral liquid chromatography involves introducing a racemic mixture into a column packed with a material that has chiral properties, commonly a modified version of amylose, a type of starch. As the mixture is propelled through the column by a solvent, one of the enantiomers will stick to the packing material more than the other and a separation can be achieved. In this case, esketamine can be separated from arketamine.

Janssen's trials showed that esketamine is effective for treatment-resistant depression and managed to get FDA and Health Canada approval for Spravato, a nasally administered version. However, there was no direct comparison with generic, racemic ketamine, and expert opinions differ on whether Spravato is worth the extra expense, which is considerable.

About 5 percent of adults suffer from depression, a leading cause of disability worldwide, and about 30 percent of cases are treatment resistant, meaning that they have not responded to at least two antidepressant medications. If Spravato does turn out to be truly effective, it would be yet another triumph for chemistry and a welcome treatment for victims of depression. Only time will tell.

CLARENCE BIRDSEYE AND TV DINNERS

The fish tasted like it had just been caught! Clarence Birdseye was amazed. He had eaten frozen foods before but they had never tasted quite right. What was the difference? This fish did not come from any commercial processing facility; it had been frozen by the Inuit fisherman right after he had caught it through a hole in the ice.

Birdseye consumed that epic meal sometime around 1914 while he was working as a surveyor for the U.S. Department of Agriculture in Labrador. The winter temperatures sometimes reached −40°Celsius in Labrador, which meant that a fish would quickly freeze after being pulled from the water. Perhaps, Birdseye thought, this rapid freezing was the key to the retention of the texture and flavor. He was correct. If food is frozen slowly, there is time for its water content to convert to large ice crystals that can damage cells and lead to off tastes and a mushy texture. Quick freezing results in much smaller ice crystals that are less disruptive.

Once Clarence Birdseye returned to the U.S., he followed up on his Labrador experience and experimented with flash freezing of fish fillets and soon developed and patented a "double belt freezer" in which cartons of fish were frozen as they passed between two refrigerated surfaces cooled by a brine solution. In 1925 he founded the General Seafood Corporation, and four years later, after further improvements of the machinery, Birdseye sold the company to Postum, which eventually became the General Foods corporation. The price was a stunning $22 million!

Birdseye of course did not invent frozen foods; the ancient Chinese already been preserving food in ice caves and American food producers had dabbled with selling frozen foods. But Birdseye's technology had made the food far more palatable. Frozen peas were marketed as being "as gloriously green as any you will see next summer," and peas were joined by spinach, fruits, berries, and meat. Frozen foods competed with the canned foods to which the public had become accustomed, but sales got a huge boost during World War II when cans were rationed. There were a couple of reasons for this. Canned foods were ideal to be sent overseas to soldiers, and metals were needed for the war effort. Tin specifically was in short supply since Japan was the largest producer of the metal and it was needed for airplane parts, ammunition boxes, solder, and especially for syrettes.

The syrette, developed by the Squibb pharmaceutical company, was a huge advance in medicine. It consisted of a small tin tube, much like a toothpaste tube, that was filled with morphine and fitted with a small hypodermic needle. Syrettes were carried by soldiers, who, if wounded, could self-administer the painkilling drug. Tin reclaimed from two cans was enough to manufacture one syrette and millions of syrettes were needed. The metal became so valuable that canned foods were rationed and canned pet foods were eliminated. Interestingly, this led to innovation with the development of dried pet food that now makes up most of the market. A huge publicity drive to salvage cans was launched with the slogan "save'em, wash'em, clean'em, squash'em."

With the need for weapons, tanks, planes, and ammunition, factories shifted from manufacturing civilian goods to military supplies. Automobile assembly lines were reconfigured, and no cars were produced between 1942 and 1945. The public was asked to donate non-essential typewriters to the military, and even sliced bread was banned because automatic slicers used metal blades.

After the war, metals again became available for consumer products, including aluminum, the production of which had been commandeered by the military because of its critical importance in producing aircraft. Aluminum is an excellent conductor of heat and reacts minimally with food, making for an ideal container when it comes to producing frozen foods. In the early 1950s airlines began to serve passengers frozen food in little aluminum trays heated in specially designed ovens instead of the usual cold sandwiches. And it was the memory of being served such frozen food, at least so the story goes, that launched the iconic TV dinner.

The Swanson company had capitalized on Birdseye's quick-freeze technology and was producing frozen turkeys on a large scale. It seems that in 1953 someone made a miscalculation and there was overproduction of some 260 tons of frozen birds. It was then that Swanson salesman Gerry Thomas recalled being served heated food in a metal tray on a business flight and thought the technology could be applied to the sales of the frozen birds. There was nothing terribly innovative about that, nor about the compartmentalized aluminum trays that allowed turkey slices to be accompanied by mashed potatoes and peas. As early as 1944 airlines had served frozen dinners called Strato-Plates on a paperboard tray coated with Bakelite resin. Then in 1950, the first aluminum tray for frozen meals was introduced as the "FrigiDinner," but it was the Swanson Company that captured the frozen dinner market with Thomas's brainchild of designing the tray in the shape of a television screen! America had been thoroughly captivated by television, and meals at the dinner table were rushed through so that the family could settle down on the couch for the evening's entertainment. Why not skip the dinner table and enjoy a full-course meal in front of the TV? thought Thomas.

Not a bad idea, as it turned out. In 1954, Swanson sold ten million TV dinners! The origin of that fruitful idea has, however, been contested by heirs to the Swanson fortune, who claim that Clarke and Gilbert Swanson, who ran the company in the 1950s, came up with the design of the iconic tray. Whoever was responsible, the fact is that the idea has stuck. Swanson frozen meals are still available, although the TV connection has been abandoned. Numerous other companies have joined the frozen food frenzy, producing a stunning variety of individual foods and meals. Freezing is an excellent method of preserving foods, requiring no added preservatives and generally providing better taste than canned food. Thanks, Clarence Birdseye! Now I'll go and pop my presliced frozen bagel in the toaster. Tastes like fresh. Well, almost.

DIAMONDS!

"O Diamond, Diamond, thou little knowest the mischief thou hast done." Isaac Newton supposedly spoke those words upon seeing his dog Diamond upset a candle and set fire to a manuscript he had been working on for twenty years. While the story of the mischievous dog is likely apocryphal, the fire was real. Newton was very interested in alchemy and had prepared an extensive manuscript on the subject. It was that work that mostly went up in flames, although parts survived. In 2020, three leaves of the scorched document sold at auction for over half a million dollars!

If Newton did have a dog, he may very well have named it Diamond because the world's first truly great scientist was interested in diamonds. As Alexander Pope wrote in his famous couplet, "Nature, and Nature's Laws lay hid in Night. God said, *Let Newton be!* and all was Light." Cleverly stated, since much of Newton's work focused on light and how it traveled from one medium to another. His classic experiment showing that white light passing through a prism can be separated into the colors of the rainbow introduced the concept of "refraction," the phenomenon of light being deflected as it passes from one medium into another.

In this connection, Newton studied light passing through a diamond and noted that it refracted in a fashion similar to "inflammable bodies" such as olive oil, turpentine, and amber. He conjectured that the diamond is "an unctuous (oily) body coagulated," thereby becoming the first person to reflect on the chemical makeup of a diamond. Whether he tried to combust a diamond, usually the first step in determining the composition of a substance, is not known, but he very well may have since he was of course interested in lenses and was undoubtedly familiar with their ability to achieve extremely high temperatures by focusing sunlight.

The brilliant French chemist Antoine Lavoisier did carry out a pioneering experiment along these lines in 1772 when he focused light on a diamond in a closed glass vessel and noted that the diamond was set aflame! He even managed to trap some of the air in which the inflammation had taken place and found that when it was passed through limewater (calcium hydroxide) it formed a precipitate of calcium carbonate. This was a giveaway that the combustion had produced carbon dioxide gas, with the carbon coming from the diamond.

Lavoisier was not able to make a quantitative connection between the weight of the diamond and the amount of carbon dioxide produced, but the English chemist Smithson Tennant managed to do exactly that in 1796. He burned a weighed sample of diamond, collected and weighed the calcium carbonate produced, and determined that the carbon content of the calcium carbonate was equal to the weight of the diamond that had burned. This meant that the diamond was composed of pure carbon!

Indeed, today we know that diamond is a crystalline form of carbon in which each carbon atom is bonded to four others in a tetrahedral arrangement, extending in three dimensions. This arrangement makes diamond the hardest known substance. Naturally occurring diamonds were formed billions of years ago deep within the earth, not from coal as was first surmised, since coal is the end product of decomposing plant or animal matter and diamonds were formed before these ever appeared. The consensus is that diamonds were formed from minerals such as various carbonates upon exposure to high temperatures and pressures in the bowels of Earth, and were brought to the surface via volcanic eruptions. Diamonds are rare, their mining laborious, polishing and cutting difficult, and demand high, which accounts for their high cost.

Due to the beauty and scarcity of diamonds, it is understandable that ever since Tennant's discovery that the gems are made of carbon, scientists have been intrigued by the possibility of producing diamonds in the lab. French chemist Henri Moissan's interest was triggered by examining rock samples from a meteorite that landed in Arizona that he believed contained tiny diamonds. He later determined that these were actually crystals of silicon carbide, a substance that was almost as hard as diamond. Nevertheless, he believed that the heat and pressure experienced by a meteorite could be reproduced in the lab and possibly produce diamonds.

Moissan developed an electric arc furnace that could reach temperatures in excess of 3,500°Celsius in which he placed a crucible made of carbon that contained iron. When molten iron is rapidly cooled, it contracts and produces great pressure, possibly enough to convert the carbon to diamond. Indeed he thought he had successfully produced a diamond, but that has been questioned. However, he did go on to make silicon carbide synthetically, and that was almost as good as making a diamond. It gave rise to moissanite jewelry, which simulates diamonds. It is far cheaper than diamond and has similar brilliance. Moissan's other interest was in minerals that contained fluorine from which he eventually isolated the element fluorine — an accomplishment that received the Nobel Prize in Chemistry in 1906. Fluorine has numerous uses such as in the production of Teflon and stain and water repellant substances known as perfluoroalkyl substances (PFAS). These are controversial because of their environmental persistence and potential toxicity.

Real synthetic diamonds were first produced in 1955 by researchers at General Electric who subjected graphite, another crystalline form of carbon, to extreme temperatures and pressures. These are real diamonds, indistinguishable from naturally occurring ones, but cheaper. They should be distinguished from the even more economical diamond simulants such as cubic zirconia, made of zirconium oxide, or moissanite, which look very much like diamonds but have a different chemical composition.

Many people prefer "real" mined diamonds and look on all others as "fake." I actually appreciate the "fakes" more, especially synthetic diamonds, given the chemical ingenuity that has gone into making them. There is also no concern about them being "blood diamonds" mined in war zones and sold to finance conflicts.

HEAD TRANSPLANTS

"The boundaries which divide Life from Death are at best shadowy and vague. Who shall say where the one ends, and where the other begins?" So wonders the unnamed narrator in Edgar Allan Poe's "The Premature Burial." At the time the classic short story appeared in 1844, the public was fascinated by cases of people mistakenly pronounced dead and buried alive. There was even a market for coffins equipped with emergency devices that would allow a reawakened "corpse" to call for help. While there is scarce evidence that premature burials ever occurred, there have been philosophical and scientific questions about the moment at which life converts to death. Is there a difference between "brain death" and "biological death"? That is, could a person be declared dead if the heart is still beating?

In 1968, a Harvard committee was tasked with defining when death occurs. The conclusion was that when a person has no detectable brain activity according to a number of criteria, then that person is dead since there has never been a case of anyone who's been correctly diagnosed as brain-dead ever showing any neurological recovery. This definition, however, has been the subject of a number of legal challenges such as in the case of Jahi McMath, a thirteen-year-old girl who had severe blood loss after tonsil and adenoid surgery that deprived her brain of oxygen. She was placed on a ventilator but was soon declared brain-dead. Her mother refused to accept that declaration because Jahi's heart was still beating normally. The hospital wanted to terminate life support and a legal battle ensued with the mother eventually finding a facility in New Jersey that accepted Jahi and kept her on a ventilator until she died five years later without ever showing any sort of brain activity.

The case raises important questions. How can someone be dead in one location and not in another? What about people who believe in the "healing power of God" and hold out hope that a miracle may occur as long as the heart is beating? Or those who believe that science will discover a way to reanimate the brain as long as the body is kept alive? There are also issues about costs and the extensive medical care required to sustain life artificially.

Then there are the philosophical questions. If humans have a soul, where does it reside, and when does it leave the body? Back in 1907, physician Duncan MacDougall designed an experiment to find out. He identified immobile patients in a nursing home who were close to death and placed their bed on a giant scale. MacDougall claimed that the patients' weight decreased by 21 grams at the time of death, which he attributed to the soul leaving the body. He followed up by carrying out the same experiment with dogs and suggested that the observation of no weight loss corroborated the notion that only humans have a soul. MacDougall's experiments were widely criticized on numerous grounds, including flawed methods, small sample size, and the likelihood that he poisoned the dogs.

Soviet physiologist Vladimir Demikhov didn't care whether dogs had souls. What they had was a heart and circulatory system similar to that of humans. In 1951 he performed the world's first heart transplant on a dog, eventually improving his pioneering methods to permit one dog to live for seven years with a transplanted heart. If hearts could be transplanted, what about heads? In 1954, Demikhov carried out the bizarre experiment that would eventually garner him worldwide attention when he transplanted the head of a puppy onto the body of a larger dog. Connecting the head's vascular system to that of the host allowed the grotesque two-headed animal to survive for days with both heads capable of moving and even eating. When questioned about his work, Demikhov would quip that "two heads are better than one."

While Demikhov's head transplant conjured up images of Victor Frankenstein, there is no question that his heart, liver, and lung transplants in animals laid the foundations for organ transplants in humans. Dr. Christiaan Barnard, who carried out the world's first heart transplant in 1967, was inspired by Demikhov, as was famed American neurosurgeon Dr. Robert White, who is remembered as much for his transplant of the head of a monkey onto the body of another as for his pioneering demonstration that cooling the brain gives surgeons more time to carry out successful operations.

Amazingly, the transplanted monkey head remained functional for eight days, and even attempted to bite Dr. White's finger, seemingly remembering his tormentor. To White this meant that the "essence" of the monkey was in the brain and led him to contemplate a human head transplant. Perhaps this would allow a brain, and possibly the soul, to survive after the body has failed. Stephen Hawking would be a perfect candidate, White believed. The head containing the brilliant brain could live on, attached to a new body after his own ceased to function. White, who died in 2010, even experimented with head transplants on corpses!

Since such transplants involve the severing of the spinal cord, the body would be paralyzed, as was the case with the monkeys. But Italian surgeon Sergio Canavero refers to some animal experiments in which spinal cords have been reconnected and has said he plans to carry out a human head transplant, probably in China. Based on what is known today, this seems like some macabre experiment by a mad scientist, but then again, once organ transplants were considered to be impossible. Now they allow for life where otherwise there would be death.

ORGANOCATALYSIS

The Nobel Prize is the epitome of recognition when it comes to scientific achievements. These annual awards were established in 1895 in Alfred Nobel's will with the aim to recognize scientists who "during the preceding year have conferred the greatest benefit to humankind."

Nobel had accumulated a considerable fortune from his invention of dynamite, an explosive that had practical uses in construction, but he was also concerned that it could be used to the detriment of humankind. Nobel envisioned that his prizes would stimulate research from which society would profit. That intent has certainly been met by the work of Benjamin List of the Max Planck Institutes in Germany and David MacMillan of Princeton University. The two scientists were awarded the 2021 Nobel Prize in Chemistry for their development of "an ingenious tool for building molecules." That ingenious tool is organocatalysis.

Life is all about building molecules. Our bodies are constantly linking amino acids to make proteins, synthesizing adenosine triphosphate (ATP) to store energy and producing neurotransmitters along with a host of other molecules needed to sustain life. In research labs and industrial facilities around the world, chemists synthesize medications, polymers, dyes, paints, agrochemicals, cosmetics, cleaning agents, disinfectants, and numerous other substances that have become an integral part of our lives. Many of these reactions rely on the use of catalysts, substances that dramatically speed up a chemical reaction without themselves being consumed. Drs. List and MacMillan independently developed "organocatalysts" that facilitate the "building of molecules" by reactions that without them would not normally proceed in a significant fashion.

The concept of catalysis is not new. Dissolve sugar in water and nothing happens. At least not at any observable rate. Add some yeast, and the sugar is converted to carbon dioxide and alcohol. That's because yeast produces zymase, an enzyme that acts as catalyst and speeds up the reaction. In our bodies, enzymes help digest fats, synthesize DNA, and eliminate toxins. Industry also uses enzymes to produce biofuels, stain removers, pharmaceuticals, and as agents to break down waste products. However, most catalysts used commercially are various metals or their derivatives. For example, one of the most important industrial processes is the reaction of nitrogen with hydrogen in the Haber-Bosch process to make ammonia, the fertilizer that has saved millions from starvation. Catalytic converters on cars, the polymerization of ethylene, the synthesis of vitamin A, and the production of drugs such as protease inhibitors to treat hepatitis C all rely on metal catalysts.

While enzymes and metal catalysts are widely used, there are some issues. Enzymes have a limited temperature and pH range, and while metallic catalysts have a wide scope, their use presents difficulties since they have to be kept free of moisture and oxygen to function properly. Furthermore, traces of metals can be left behind in an unacceptable fashion when it comes to producing medications or food.

Drs. List and MacMillan developed simple organic molecules, termed "organocatalysts," that do not contain any metal atoms and also have some novel capabilities. List wondered how enzymes, basically long strings of amino acids, enhance reactions. He determined that it was one particular amino acid in the chain, proline, that attracted the reagents and facilitated their engagement. This led to trying just pure proline as a catalyst, and it worked! MacMillan, who actually coined the term "organocatalysis," independently discovered that another small organic molecule, imidazolidinone, also catalyzed a number of reactions. A descendent of this has been christened "MacMillan catalyst" and is widely used.

Furthermore, organocatalysts also had the ability to carry out asymmetric synthesis. This comes into play with some molecules that can exist in non-superimposable mirror-image forms, like our hands. That possibility was actually demonstrated by Jacobus van't Hoff, who in 1901 was the first-ever recipient of the Nobel Prize for Chemistry. This molecular feature can be particularly important with the synthesis of pharmaceuticals that involve many steps. Some of these steps often yield intermediates that can exist in such dual forms, termed "enantiomers." This can be a problem since one of the mirror image forms can lead to the desired product, while the other is an undesired, sometimes toxic, contaminant. Organocatalysts can be used to synthesize only the desired version, important in the production of medications such as the antico-agulant coumadin, the cancer drug paclitaxel (Taxol), the antidepressant paroxetine (Paxil), and the antiviral oseltamivir (Tamiflu).

Another area where organocatalysts have made a mark is in the pursuit of "green chemistry," which has the goal of designing chemicals, chemical processes, and commercial products in a way that avoids the creation of toxins and waste materials. For example, polymerization reactions to produce polystyrene and polyvinyl chloride traditionally have required high temperatures and the use of expensive metal catalysts that also come with the baggage of environmental toxicity. Organocatalysts are cheap, non-toxic, and work at a lower temperature, saving energy.

It has been estimated that about 35 percent of all the finished goods and services in the world rely on the use of catalysts. There is no question that the discoveries of Drs. List and MacMillan, first published in 2000, have already impacted our lives and will do so even more in the future.

THE BARK THAT CURES

The physicians in the court of King Charles II were apoplectic. The king had appointed Robert Talbor, a man they considered to be an unqualified charlatan, as his personal physician. Indeed, Talbor had no training as a physician, although truth be told, whatever training doctors had at the time was all about purging, bloodletting, and various herbs aimed at restoring the balance of the four bodily humors: blood, yellow bile, black bile, and phlegm. The humoral theory that had been championed by Hippocrates and Galen held sway for some 2,000 years despite the lack of any scientific validity.

Tabor had been an apprentice to an apothecary in Cambridge where he had learned about the Jesuits having introduced a medicinal tree bark from South America to Europe around 1630. One story, almost certainly apocryphal, is told of the Countess of Chinchon, wife of the Spanish viceroy of Peru, being cured of what at the time was called "tertian fever" with a preparation made from this special bark. Tertian fever was so-called because the fever would cycle roughly every three days. Today we know it as malaria. The countess immediately ordered that the bark be given to the sick of Peru and sang its praises when she returned to Spain. This is when Jesuit Cardinal de Lugo heard about the bark and took it for testing to Rome, from where "Jesuit bark" spread throughout Europe. One of the problems with the romanticized story is that the countess never returned to Spain.

An account that has more historical evidence describes South American Indigenous people who had to cross a river up to their necks in cold water finding a remedy to stop their shivering. They would drink a decoction of the bark in hot water. Jesuit missionaries learned of this practice and, reasoning by analogy, tried it for malaria and found that it worked. They then introduced it to Europe as well. There is a problem with this story as well, since quinine, the active ingredient in the "Jesuit bark," works against malaria by killing the parasite that causes the disease and would not be effective to stop shivering from cold. In any case, the Swedish botanist Linnaeus seems to have believed the Countess of Chinchon fable and named the tree "cinchona."

The use of cinchona bark was mired in controversy. Physicians in general questioned its use since it did not have any purgative effect and therefore did not fit into the humoral theory of disease. Also, since there was no standardized way of administering the bark or its extracts, the remedy did not always work. Then along came Talbor, who in 1672 introduced a "secret remedy" against malaria in his book *Pyretologia, A Rational Account of the Cause and Cure of Agues*, in which he warned about problems that could befall sufferers who were treated with Jesuit bark. The kicker here is that the "secret remedy" was in fact cinchona bark, and Talbor had found a way to prepare a reliable extract. A French nobleman who had landed in Essex on his way to discuss battle plans against the Dutch with King Charles came down with malaria and heard about Talbor's remedy. He was so impressed with the way he was cured that he recounted the experience to the king, who immediately sent for Talbor and was so taken with the man that he appointed him his personal physician, drawing the outrage and fierce criticism from the College of Physicians.

When the son of Charles II's cousin, Louis XIV of France, became ill with malaria, Charles dispensed Talbor to help. The boy was cured, and so was the Queen of Spain, who had also contracted the disease. Louis was so impressed that he offered Talbor a large sum to reveal the secret of the cure, to which he agreed as long as it would not be during his lifetime. When Talbor died at the age of forty, a wealthy man, Louis commissioned a book in which the secret was revealed to be cinchona bark steeped in rose leaves, lemon juice, and wine. The remedy became popular until it was superseded by a preparation of almost pure quinine that Pierre-Joseph Pelletier and Joseph-Bienaime Caventou had managed to isolate in 1820, initiating the large-scale manufacture of quinine and saving multitudes from the misery of malaria.

While Talbor had some elements of a charlatan with his insinuation of having found a "secret formula," Charles II's faith in the man paid off. Not only would Talbor's product cure many in Europe, Charles himself would directly benefit when he came down with malaria. It is interesting to note that Charles was a great proponent of science, having been tutored as a young man by William Harvey, the surgeon who had first described the circulatory system. As king, Charles became acquainted with the work of Christopher Wren, Robert Hooke, and Robert Boyle, and in 1660 awarded a charter to the Royal Society, thereby establishing an organization that has been promoting excellence in science for the benefit of humanity for over three centuries. Charles himself was greatly interested in science and even had a private chemistry lab. Unfortunately, some of his experiments involved distilling mercury, which could have contributed to his death, which has been theorized to have been due to irreversible kidney disease possibly caused by mercury poisoning.

In his final days Charles was subjected to bloodletting, purging, and cupping, all useless, torturous treatments administered by doctors who had called Talbor a quack for using a remedy that actually worked.

SCHO-KA-KOLA

The Coca-Cola Company was not happy. In 1999, a German chocolate manufacturer had filed an application in the U.S. to register Scho-Ka-Kola as a trademark for its brand of caffeine-rich chocolate produced from cocoa beans, coffee beans, and the fruit of the kola tree. Coca-Cola disputed the registration on grounds that "the name is likely to cause confusion with, and dilute, the famous trademark 'Coca-Cola,' which has long been used and registered for beverages and a wide range of products." The dispute was upheld and registration of Scho-Ka-Kola as a trademark was denied.

Chocolatier Theodor Hildebrand first formulated Scho-Ka-Kola in 1935 with the idea of producing an ideal stimulant for German athletes at the upcoming Berlin Olympic games. The popularity of the brand increased dramatically during the war when the chocolate was provided to Luftwaffe pilots as well as to tank and submarine crews to induce wakefulness and alertness. This led to the myth, repeated in many historical accounts, that the real reason that the chocolates, known colloquially as "Fliegerschokolade," or "Aviator chocolate," were so prized was that they actually contained amphetamines. They did not. The stimulant effect was due completely to caffeine, with a serving containing about as much caffeine as a strong cup of coffee.

While Fliegerschokolade never contained amphetamines, these drugs were widely used during World War II both by the Germans and the Allies. Amphetamine was first synthesized in 1887 by Romanian chemist Lazar Edeleanu, who was looking for an improved version of ephedrine, a naturally occurring component of the ephedra plant that had been isolated just two years earlier. "Ma huang," as ephedra is known in Traditional Chinese Medicine, was of interest because of its rich history as a stimulant and an aid for respiratory problems.

As is often the case, when a plant component has medicinal value, chemists explore the possibility of making changes in the basic molecular structure for greater efficacy. This is exactly what Dr. Edeleanu was attempting to do when he synthesized amphetamine. However, his interest switched to developing a process for refining crude oil, and he failed to pursue amphetamine further. In 1932, American chemist Gordon Alles, unaware of Edeleanu's earlier work, independently synthesized amphetamine, again trying to improve on the action of ephedrine. He first tested the new compound on guinea pigs and then became his own guinea, noting that his nasal congestion disappeared. He also experienced a "feeling of well-being!" Dr. Alles approached the Philadelphia pharmaceutical firm Smith, Kline & French about a partnership, and soon Benzedrine entered the marketplace as a treatment for congestion and asthma in the form of an inhaler. It wasn't long before the drug attained a reputation as a stimulant, especially after its use by American athletes at the Berlin Olympics.

German chemist Friedrich Hauschild at the Temmler-Werke pharmaceutical company was aware of the use of Benzedrine at the Olympics, and trying for one-upmanship, synthesized methamphetamine, a close cousin of amphetamine. This compound had actually first been made from ephedrine in 1919 by Akira Ogata in Japan, but Hauschild developed a method to produce the drug on a large scale under the name Pervitin.

Although Pervitin was available to the general public in pharmacies, it was on the battlefield that it would make its mark. Nazi ideology considered the use of social drugs as a sign of weakness and moral decay, but Pervitin was an exception. Unlike alcohol or opiates, methamphetamine was not considered to be about escapist pleasure but rather about achieving physical and mental superiority, very much in line with Nazi goals. Methamphetamine-enhanced soldiers would require less sleep and fight longer and harder!

This was exactly what was needed for Blitzkrieg, a lightning-quick strike that would catch the enemy off-guard. When German troops invaded Poland, they were energized with Pervitin, but the drug probably played its biggest role in the invasion of France through the Ardennes forest. The Allies had assumed that because of the challenging terrain the German advance would be slow and there would be time to move defensive troops into position. But General Heinz Guderian, who led the invasion, demanded that his tank crews go sleepless for at least three nights to speed the advance, and Pervitin made that possible. In his memoirs, Churchill noted that he had been dumbfounded by the German tanks advancing night and day.

With the extensive use of Pervitin, problems began to crop up. There were reports of high blood pressure, heart attacks, and addiction. By 1941, the Germans had cut back on the use of methamphetamine, but Benzedrine became a staple for British and American troops to combat fatigue and boost morale. In Japan, where the drug was produced as "Philopon," it was commonly issued to kamikaze pilots. Later, amphetamines were widely used by the U.S. military in the Korean, Vietnamese, and Persian Gulf wars to decrease fatigue.

Today, methamphetamine is involved in a different kind of warfare. The fight is between authorities trying to curb its illegal production by clandestine labs that crank out billions of tablets of crystal meth, a form of the drug that can be smoked. Rising crime rates and a host of medical problems are the price paid when it comes to the quest for instant euphoria by individuals addicted to crystal meth.

As far as Scho-Ka-Kola goes, it is still produced and is popular in Germany, commonly sold at gas stations to keep drivers alert. Although not available in stores here, like almost everything else, it can be purchased online. But it is much more expensive than a cup of coffee.

EXPANDING ON SPANDEX

In 1922, Johnny Weissmuller, who would go on to fame portraying Tarzan in the movies, stunned the sporting world by swimming the 100-meter freestyle in under one minute with a time of 58.6 seconds. Nobody cared or noted what kind of swimsuit he wore. It was simple cotton. Quite a contrast with the high-tech suit worn by American Caeleb Dressel, who took the gold medal in the event at the Tokyo Olympics with a time of 47.02 seconds!

Of course, in the intervening 100 years, training methods have changed, although Weissmuller did place emphasis on lifestyle. He became an enthusiastic follower of Dr. John Harvey Kellogg's vegetarian diets, enemas, and exercise. Dressel is not a vegetarian, loves meatloaf, and starts his day with a high-carb breakfast. The real difference is in the training. In addition to swimming, Dressel trains on a rowing machine and a stationary bike with online interactive personal training. But his swimsuit unquestionably also makes a difference. Certainly not ten seconds' worth, but when today's top swimmers are separated by fractions of a second, the fabric and style of the swimsuit take on importance.

Any discussion of swimsuit technology has to start with the wonders of spandex, a synthetic material that stretches and magically rebounds to its original shape like rubber. But unlike rubber, it can be produced in the form of fibers that can be woven into a fabric. Spandex, a clever anagram of "expands," was developed in the 1950s by DuPont chemist Joseph Shivers working under the direction of William Charch, who had become famous for inventing waterproof cellophane by coating the material with a layer of nitrocellulose. Revolutionizing sportswear was not Shivers's original intent. At the time, girdles made with rubber were a common part of women's attire, but rubber was in short supply and the challenge was to develop a synthetic material that could be used in girdles as a replacement.

DuPont already had introduced polymers such as nylon and poly-ester to the marketplace and had significant expertise in synthesizing giant molecules. Shivers produced spandex by synthesizing a block-copolymer with alternating elastic and rigid fragments. There were also branches that could be used to cross-link the molecules, confer-ring strength. Combining spandex with cotton, linen, nylon, or wool resulted in a material that was stretchy and comfortable to wear. Since a number of companies began to produce such fabrics, DuPont patented the name "Lycra" for its version of spandex.

In 1973, East German swimmers sported spandex suits for the first time and shattered records. That may have had more to do with their use of steroids, but it got the competitive gears turning at Speedo. The company had been established in 1928 as a science-based swim-suit manufacturer, replacing cotton with silk in its racerback suit to cut down on drag. Now, spurred by the success of the East Germans, Speedo turned to coating spandex with Teflon and contoured the surface to have tiny v-shaped ridges like those found on the skin of sharks that supposedly reduce turbulence.

By 2000, this had morphed into a full-body suit that further reduced drag, since water was found to adhere more strongly to skin than to the swimsuit material. In 2008, strategically placed polyure-thane panels replaced Teflon, and the fabric, now composed of Lycra, nylon, and polyurethane, was found to trap tiny air pockets that buoyed the swimmer. The advantage here is that air resistance is less than water resistance. Some companies tried suits of pure polyure-thane since the material traps air very effectively. With each of these "breakthroughs," times dropped and prices increased. A high-tech suit could now cost over $500.

The term "technological doping" invaded our vocabulary, and in 2009, the international governing body of swimming (FINA) decided to equalize the field and banned all full-body swimsuits as well as any panels not made of woven fabric. That didn't stop the race for improved suits even though they were now restricted in the amount of body surface they could cover. For the Tokyo Olympics, Speedo introduced yet another innovative suit that was constructed of three different layers of fabrics, the identity of which is propriety information.

Spandex is not restricted to swimwear. Skiers reduce air drag by squeezing into sleek spandex suits, as do bicyclists. Women's undergarments still make up a large portion of the business, and spandex has even made it into leggings and jeans that squeeze the body in just the right places to hide undesired bulges. As far as the swimming innovation goes, maybe competitors will just spray their naked body with some sort of polymer to eliminate any swimsuit drag! After all, the original Olympians competed in the nude.

SWAROVSKI CRYSTALS

The stage for the 2018 Academy Awards at Hollywood's Dolby Theatre sparkled as if adorned with diamonds. But there were no diamonds in sight, save for the ones that twinkled on the movie stars in the audience. The glittering reflections from the stage were the result of 45 million brilliant Swarovski crystals weighing some 15,000 pounds that had been painstakingly mounted on the proscenium! Actually, they weren't really crystals, even though that term is commonly used to describe pieces of glass that are specially cut so that their surface features numerous angled facets that reflect light.

Real crystals are composed of atoms, molecules, or ions arranged in an ordered pattern that extends in three dimensions. Salt, sugar, emeralds, iron, ice, and diamond are true crystals. Glasses on the other hand are "amorphous" (from the Greek "without shape") substances that lack an ordered composition.

In all likelihood, volcanoes produced the first samples of glass that humans ever encountered. Obsidian is a dark glass that forms when silicate minerals in lava cool and solidify. The first effective cutting tools were probably made of obsidian. Knowing where this substance was found, early humans would have naturally tried to simulate nature and attempt to heat various minerals and then cool them to produce glass. They likely tried all sorts of substances and eventually discovered that sand was the best candidate.

Pliny, the Roman historian, recorded his version of the story in the first century AD, but given that he speaks of events that happened some 4,000 years earlier, his account may not be exactly reliable. In any case, Pliny maintains that the Phoenicians were the first to produce glass when they built fires on a beach and supported their pots on chunks of natron, or sodium carbonate. The sand and natron fused under the high heat and yielded glass on cooling. Natron acts as a "flux," lowering the melting point of the sand. Pliny's account is probably apocryphal, but we do know that by about 1500 BC the Egyptians were making glass bottles. They found that adding limestone, or calcium carbonate, strengthened the glass and made it water resistant. And that basically is still the process used today. Sand, soda (sodium carbonate), and limestone are melted together, and the mix is cooled to form "soda lime glass."

In the seventeenth century, Englishman George Ravenscroft added lead oxide to the mixture and produced "lead glass" that was more brilliant, easier to melt, and more suitable for blowing into molds. This is the type of glass that young Daniel Swarovski learned to cut in the 1880s in his father's small workshop in Bohemia, now the Czech Republic. After attending an electrical exhibition in Paris, Swarovski developed and patented an electric cutting machine that reduced costs and made the production of diamond-like "crystal jewelry" possible at an affordable price. In 1895 Swarovski established a crystal-cutting factory in Austria with a vision of making "a diamond for everyone." He was always clear about his products being made of glass and only resembling diamonds. However, it is the specific way the glass is formulated, cut, and ground that gives the jewelry its brilliance. Obviously a closely guarded trade secret. Today, the Swarovski company has over 30,000 employees and an annual revenue of over $4 billion!

Swarovski crystals first appeared on the silver screen in 1932, when they dotted Marlene Dietrich's costumes in *Blonde Venus*. In 1939's *Gone With the Wind*, Vivien Leigh swooshed about in a Swarovski-speckled gown, and in *The Wizard of Oz*, the world's most famous shoes, Dorothy's ruby red slippers, dazzled with Swarovski crystals. When Marylin Monroe sang "Diamonds Are a Girl's Best Friend" in *Gentlemen Prefer Blondes*, she wasn't wearing diamonds but an assortment of Swarovski jewelry. The gown Marilyn wore when she famously serenaded President Kennedy with her breathy rendition of "Happy Birthday" at Madison Square Garden was decorated with 2,500 Swarovski crystals. Audrey Hepburn's tiara in *Breakfast at Tiffany's* sported the crystals, as did Michael Jackson's famous glove. Liberace pranced about in Swarovski-studded costumes, and the stones also decorated Elvis's jumpsuits. The famous chandelier in the film version of *The Phantom of the Opera* featured $1.2 million worth of Swarovski crystals.

While it was the clear, diamond-like brilliance that first made Swarovski jewelry famous, today's crystals can bedazzle with color. The addition of cobalt oxide to the melt produces blue, chromium oxide yields green, cadmium sulfide results in yellow, and tiny particles of gold give a ruby-red appearance. In 2012, the company announced a major change in its manufacturing process with the elimination of lead. Although lead is a nasty toxin, there was never any risk to the consumer posed by Swarovski crystals. The change was made to avoid any perception of a hazard in light of the extensive publicity given to the dangers of lead-based paints and to water contaminated with lead. The replacement for lead oxide is barium oxide, which gives comparable brilliance. It is quite easy to tell if a crystal is made with lead or barium, since the barium version will be considerably lighter. Today, Swarovski has expanded to producing lab-grown diamonds that are identical to mined diamonds.

Unfortunately, the Swarovski history is not quite as brilliant as the products the company produces. In the 1930s, members of the family were enthusiastic supporters of the Nazi party, with Daniel's son Alfred praising Hitler at business gatherings, and even donating money towards building a holiday home for the Führer. That takes away a bit of the glitter from the spectacular set of miniature Swarovski ducks I once received as a gift.

CATALYTIC CONVERTERS AND CRIME

Dr. William Hyde Wollaston did not like practicing medicine. In 1797, he declared he felt like a slave to the profession and decided to "turn his time to a less irksome" pursuit. Luckily for the world, that pursuit turned out to be chemistry. Thanks to a large inheritance from a brother, Wollaston was able to set up a lab dedicated to his pet project of refining platinum ore into pure ingots of the metal, which he managed to do through a complex sequence of reactions, becoming the first person to market the pure metal.

During the purification process, in 1803, Wollaston discovered that the platinum ore contained traces of two other metals not previously noted. He managed to purify these as well, naming one "palladium," curiously after Pallas, a recently discovered asteroid, and the other "rhodium" from the Greek for "rose," since the rhodium-containing precipitate from which the metal was finally isolated by reaction with zinc was of a "rosy" color. At the time, rhodium was a mere curiosity, and Wollaston could not have imagined that some 200 years later it would become the most precious of all metals, worth ten times as much as gold!

Why does rhodium cost more than $500 a gram? Because it is one of the rarest elements in Earth's crust, is difficult to isolate, and is in high demand by the automobile industry. The metal is an essential component of catalytic converters, the devices attached to the tailpipe of gasoline-fueled vehicles designed to cut down on pollution. Gasoline is a complex mix of hydrocarbons, compounds composed essentially of carbon and hydrogen. When this mixture is ignited by a spark in the engine's cylinder, gasoline produces a large volume of gases that quickly expand and push down on the piston that turns the crankshaft, which then turns the wheels. Once the gases have done their job, they are vented through the exhaust.

The burning hydrocarbons yield mostly carbon dioxide along with some carbon monoxide. Carbon dioxide is the notorious "greenhouse gas" implicated in global warming with roughly 20 percent of total carbon dioxide release coming from road transport. Catalytic converters cannot reduce carbon dioxide emission, but they can eliminate carbon monoxide, which is a highly toxic gas, by converting it into carbon dioxide. That conversion relies on the catalytic activity of certain platinum and palladium compounds on a ceramic support. However, carbon dioxide and monoxide are not the only gases produced in an internal combustion engine. Air is composed of about 80 percent nitrogen and 19 percent oxygen, and under the influence of heat generated by the engine, these gases can combine to form nitric oxide and nitrogen dioxide.

Both these gases are major pollutants. On exposure to sunlight, they react to yield nitric acid, a major component of acid rain, as well as ozone, a component of smog. Nitrogen oxides can also irritate the eyes, nose, and throat, and can even cause shortness of breath. This is where rhodium comes in. When these gases pass over a catalyst formulated with the metal, they are converted into innocuous nitrogen.

A catalytic converter contains only a couple of grams of rhodium compounds, but that is enough to make these devices attractive to thieves who then sell them to recyclers or junkyards that unethically accept them. With the increase in such thefts, authorities are putting the squeeze on dealers willing to buy used converters. Unfortunately, catalytic converters are rather easy to remove. A thief crawls under a car with a saw and within minutes he is off with the device. Should he crawl under an electric car, he'll be disappointed, since these do not burn gasoline and therefore do not require a catalytic converter.

The catalytic activity of rhodium compounds is put to use in other contexts as well. Menthol is produced in large amounts as a component of lip balms, cough medicines, decongestants, after-shaves, mouthwashes, toothpastes, candies, chewing gum, cigarettes, and perfumes. It can either be isolated from peppermint oil, or produced synthetically from myrcene, found in various plants. The conversion of myrcene to menthol is made possible by a rhodium catalyst. So is the synthesis of L-DOPA, the drug used to treat Parkinson's disease.

In 2001, Stanley Knowles, a researcher at Monsanto, at the time a pharmaceutical company, received the Nobel Prize in Chemistry for his discovery of "asymmetric synthesis" using a rhodium catalyst. Some molecules can exist in two forms that are mirror images of each other but have different physiological activities. This is the case with L-DOPA, the so-called "left-handed" form being far more active. Before Knowles's asymmetric synthesis, this was difficult to produce. Now Parkinson's patients can thank Knowles and his rhodium catalyst for improving their condition.

Heart disease patients with implanted pacemakers also benefit from rhodium. Like its chemical relatives, platinum and palladium, it is an ideal metal for such internal devices because it does not corrode and is not rejected by the body.

Finally, rhodium can be used to electroplate jewelry. An extremely thin layer can make new pieces extra bright and shiny, although eventually the shine wears off. But the rhodium coating will not wear off the record given to Paul McCartney in 1979 by *The Guinness Book of Records* in recognition of being the all-time bestselling singer-songwriter. Rhodium was chosen to indicate that this achievement deserved more than gold or platinum!

FILL 'ER UP — WITH HYDROGEN

Queen Victoria watched with amazement as Dr. John Henry Pepper (not of soft drink fame), picked up a seemingly empty bottle and proclaimed: "And now, the oxygen and the hydrogen will have the honor of combining before Your Majesty!"

With that, he pulled out the stopper and pointed the neck at an open flame. The queen of England and her entourage were astounded by a loud bang and a flash. Indeed, the hydrogen that had filled the bottle combined with oxygen in the air in a spectacular fashion! It was the 1850s, and Dr. Pepper, director of the Royal Polytechnic Institution in London, was explaining to the monarch that the two elements reacted to form water with the release of a great deal of energy. Hydrogen could turn out to be a great fuel, Pepper went on, if only it could be obtained more easily. Alas, at the time, this was not possible. Pepper made hydrogen by the method first described in 1671 by Robert Boyle, widely regarded as one of the fathers of modern chemistry. Boyle had described how the addition of acids to metal filings "belched up copious and stinking fumes which would readily take fire and burn with more strength than one would easily suspect." Since hydrogen has no smell, the stench was likely due to sulfur impurities in the metal filings generating hydrogen sulfide, the odor of rotten eggs.

Dr. Pepper wasn't the only one in Victorian times to contemplate using hydrogen as a source of energy. Jules Verne, in his classic 1874 novel *Mysterious Island*, had his shipwrecked engineer hero speculate that "water will one day be employed as fuel, and the hydrogen and oxygen which constitute it, used singly or together, will furnish an inexhaustible source of heat and light."

Although not quite there yet, we are getting closer to Verne's futuristic vision. Hydrogen is the ultimate clean fuel since the only product produced when it burns is water. Japan, aiming towards a "hydrogen society" that minimizes fossil fuels, symbolically chose hydrogen to create the flame in the Olympic cauldron. When hydrogen burns, electron-hungry oxygen atoms snatch electrons from hydrogen, with the resulting negative oxygen and positive hydrogen ions then combining to form water with the liberation of a huge amount of heat energy. However, hydrogen does not have to be burned to produce energy. It can be used to produce electrical energy in a fuel cell, a device that allows hydrogen and oxygen to combine without combustion.

A fuel cell consists of a negative electrode (anode) and a positive electrode (cathode) sandwiched around an electrolyte, a substance through which ions can readily travel. Hydrogen is piped into the anode and oxygen into the cathode. At the anode, a platinum catalyst causes hydrogen to be split into positive hydrogen ions and negatively charged electrons. These electrons then flow through an external circuit, creating a current, before being taken up by oxygen at the cathode to yield negatively charged oxygen ions that then combine with the hydrogen ions that have traveled through the electrolyte to form water. Unlike batteries, fuel cells do not die. As long as hydrogen and oxygen are available, current is generated and can be used to run an electric motor that can power a car, bus, train, or even an airplane.

There is, however, a fly in the hydrogen ointment. Today, the vast majority of hydrogen, most of which is destined for the synthesis of ammonia fertilizer, is produced through "steam reformation." In this process natural gas (methane) is reacted with water to yield hydrogen, but the problem is that carbon dioxide, the notorious greenhouse gas, is also formed. While carbon dioxide can be captured and sequestered underground, the cost is prohibitive. That is why the Holy Grail of energy production is "green hydrogen," produced without the use of any fossil fuel.

As most high school students learn, or should learn, passing an electric current through water causes it to break down into hydrogen and oxygen. "Electrolysis," a term coined by Michael Faraday, was discovered by William Nicholson and Anthony Carlisle in 1800, shortly after Alessandro Volta had introduced the "voltaic pile," essentially the first battery. While attempting to replicate Volta's experiment, the English chemists accidentally contacted the pile's wires with water and observed the formation of gases that turned out to be oxygen and hydrogen. This led to the birth of electrochemistry, the branch of chemistry that deals with chemical changes produced by electricity.

Green hydrogen can be made by electrolysis if the electricity needed can be obtained without burning fossil fuels. This means solar, wind, tidal, or hydropower. Japan has already built the world's largest solar-powered electrolyser for hydrogen production, and other countries, Australia in particular, are following suit.

The great advantage of hydrogen is that it can be stored and transported, as is required for use in fuel cells. What about safety? Hydrogen is highly combustible and discussions of its use often bring up memories of the 1937 *Hindenburg* disaster in which thirty-five people were killed when the airship burst into flames as it was landing in New Jersey. Modern hydrogen tanks made with carbon fiber are essentially explosion proof, and even if there is a leak, the lightness of the gas allows it to dissipate quickly in the air.

While hydrogen is not yet set to replace fossil fuels globally, it is making significant inroads. At the 2021 Tokyo Olympics, athletes were transported in hydrogen-fueled buses, and the athletes' village is being converted into the world's biggest hydrogen-powered neighborhood including the construction of a new hydrogen power station to be completed by 2024. Saudi Arabia is constructing a futuristic city, Neom, designed for a million people, that will be powered totally by green hydrogen. As far as Canada goes, plans are to have over 5 million hydrogen vehicles on the road by 2050. Jules Verne may yet turn out to be right. After all, he did correctly envision electric submarines, automobiles, and spaceflight.

THE BATTLE AGAINST FRIZZY HAIR

It all started in Rio de Janeiro in 2003. Women began to flood beauty parlors after word spread about a new treatment that claimed to straighten hair, reduce frizziness, add shine, and produce a silky smoothness. "Brazilian keratin treatment" soon became a rage and spread around the world. Before long, though, it became mired in controversy.

Not because it didn't deliver the goods, but because of what it also delivered. A dose of formaldehyde, a known carcinogen.

Understanding the science behind the treatment requires a quick course in hair chemistry. So here we go! Hair is composed of a type of protein called keratin formed by cells called keratinocytes in the hair bulb that is rooted in the hair follicle, a cavity in the epidermis, the outer layer of the skin. As the hair grows, the cells fill with keratin and die so that the hair shaft becomes basically a network of protein molecules. Genetics dictates the specific fashion in which keratin molecules assemble into three-dimensional structures, and it is this structure that determines if an individual's hair will be curly or straight.

Proteins are chains of amino acids that can be coiled in various ways. Keratin takes the shape of a helix, with the shape being maintained by "hydrogen bonds," a weak attraction between oxygen and hydrogen atoms in adjacent coils. To complicate matters, these coiled keratin helices are twisted into different shapes as a result of cysteine, one of the amino acids in keratin, binding to another cysteine fragment in a different part of the chain. Specifically, it is the sulfur atoms in cysteine that form sulfur-sulfur bridges. Changing the shape of the hair requires disruption of the various bonds responsible for the keratin structure. With these bonds broken, the chain can move around more freely in response to the stresses created by combing or the placement of curlers. If at this point the bonds responsible for maintaining the structure of keratin can be reformed, the keratin, and hence the hair fibers, will have been permanently reshaped. New hair growth will be unaffected.

Hydrogen bonds are easily broken just by exposure to water. That is why wet hair can be readily shaped. Heat, such as with a hair iron, will cause the water to evaporate and allow the hydrogen bonds to reform, keeping the hair in its new shape until moisture intervenes. To have the shape be altered permanently, the sulfur-sulfur bonds have to be broken and then reformed after the keratin molecules have been reconfigured. The chemical that has traditionally been used to break these bonds is the rather unpleasant smelling thioglycolic acid. Linking of the sulfur atoms in their new position is brought about with hydrogen peroxide, an oxidizing agent. In the hands of experts, results are generally good, but control of bond breakage and bond formation is not easy and timing is critical. Too long exposure to the chemicals can damage hair, and too short can yield unsatisfactory results.

Brazilian keratin treatment straightens hair without the damage that can be caused by permanent treatments. Wet hair is combed straight and is then infused with a mixture of formaldehyde and short chains of amino acids called peptides, often derived from keratin in sheep's wool. Formaldehyde forms a bond between the keratin and the added peptides preventing the keratin molecules from returning to their original shape. Furthermore, the realigned keratin filaments reflect light very efficiently, producing brighter, shinier hair.

So, what is the controversy? Formaldehyde is a known carcinogen and a respiratory irritant. The carcinogenicity is a legitimate concern for hairdressers, who can have frequent exposure, but is unlikely to be an issue for their customers.

In response to such worries, a number of "formaldehyde-free" keratin treatments have been introduced. Sometimes the promotion of these is simply dishonest. "Methylene glycol" avoids the use of the term formaldehyde, but this is just a solution of formaldehyde in water. "Methanal" and "formic aldehyde" are alternate names for formaldehyde. Then there are treatments that really do not contain formaldehyde, with intriguing names such as "hair Botox" or "nanoplastia." While there is some science here, it is often drowned in hype.

In fact there is no actual Botox involved, and the term is used to conjure up an impression of smoothness. "Nanoplastia" is an invented, meaningless term meant to infer some type of breakthrough technology. Usually, the products used are based on hydrolyzed keratin or collagen and some chemical other than formaldehyde that binds these to the hair. Glyoxylic acid or glyoxyloyl carbocysteine can do this, and while the term formaldehyde is avoided, these chemicals can, with the heat applied during the procedure, break down and yield formaldehyde, although amounts are likely insignificant. In 2017 researchers found that specific peptide sequences that incorporate cysteine residues can bind to keratin without the need for a binding agent. These peptides can be readily manufactured and have the potential to straighten hair without the use of "harsh" chemicals.

Another interesting technology uses the aminopropyltriethoxysilane family of products that can be incorporated into hair and then harden on contact with water. With trade names such as Filloxane, Intra-Cylane, and Fibra-Cylane, these chemicals claim to add volume to hair, reduce frizz, and hold the shape of styled hair.

With the importance that people place on the appearance of their hair, it is understandable that there is great competition among products with each one trying to carve out a niche with some inventive wording. Terms such as nourishes, replenishes, redensifies, restores, reconstructs, and rebalances are all meaningless when applied to hair, as are "organic" or "natural." "Targets the hair's DNA" is pure nonsense. Some products are promoted as "chemical-free," and should be avoided if only to send a message to companies about the use of such a ludicrous expression that is an affront to science.

THE CURSE OF MISINFORMATION

Actor Woody Harrelson cheers on claims that 5G technology is somehow linked to COVID-19. Steve Bannon, hardly a model of healthy living, recommends warding off viral infections by using a nebulizer two or three times a day with a mix of saline and hydrogen peroxide, swallowing a third teaspoon of pink Himalayan salt, and downing large doses of zinc, Vitamin D, and probiotics. Then there is Gwyneth Paltrow, who treats her lingering COVID-19 symptoms with a "keto and plant-based" diet, involving fasting until 11 a.m. every day, "lots of coconut aminos," and sugar-free kombucha and kimchi, along with Madame Ovary supplements, which, surprise, surprise, she sells. It should not come as a shock that there is no evidence for any of this purported wisdom since celebrity status does not confer scientific know-how. But when properly educated physicians spread unscientific drivel, that is a different story. One with potentially dangerous consequences.

Let's start with osteopathic physician Dr. Joe Mercola, honored by the non-profit Center for Countering Digital Hate as the number-one misinformant. Mercola has a long history of confrontation with the FDA and has received numerous warning letters for selling products without supporting evidence. The latest one directed him to "immediately cease the sale of unapproved and unauthorized products for the mitigation, prevention, treatment, diagnosis, or cure of COVID-19," citing false claims made about Mercola's Liposomal Vitamin C, Liposomal Vitamin D3, and Quercetin and Pterostilbene Advanced.

In the past, Dr. Mercola has received similar warnings about his Tropical Traditions Virgin Coconut Oil, that claimed to benefit Crohn's patients and reduce the risk of heart disease. He was also taken to task about his claim that Vibrant Health Research Chlorella XP, "helps to virtually eliminate your risk of developing cancer in the future." Mercola has also claimed that his tanning beds reduce the risk of cancer and that metal coils in mattresses "actually act like an antenna attracting and amplifying whatever radiation might be zipping through your bedroom." All this pales in face of his recent crusade against COVID vaccines.

Mercola dredges up comments to support his attack on the jab from a number of other doctors who have been widely castigated by mainstream scientists. Dr. Vladimir Zelenko's claim of having successfully treated thousands of COVID-19 patients using hydroxychloroquine (HCQ), azithromycin, and zinc sulfate has been widely disputed. This brilliant savant proclaims that there is a "very distinct possibility that everyone who receives the COVID jab may die from complications in the next two to three years."

Dr. Richard Fleming, nonsensically labeled as the "father of modern nuclear cardiology," has been convicted of two felonies under federal law for fraud, but nevertheless is another beacon of truth for Mercola.

Fleming believes the vaccines are a bioweapon that leads to an increase in Alzheimer symptoms and causes "prion" disease. Prions are the type of protein that can trigger normal proteins in the brain to fold abnormally and were found to be the cause of mad cow disease. There is zero evidence that vaccines have anything to do with prions.

Dr. Sherri Tenpenny is another osteopathic doctor revered by Mercola. She believes that vaccinated people infect others but "they don't get COVID symptoms that we typically recognize as COVID but they get bleeding, they get blood clots, they get headaches, they get heart disease." Tenpenny, drenched in conspiracy theories, promotes the idea that microchips in vaccines communicate with 5G cell towers and that vaccines magnetize people so that they can "stick a key on their forehead or spoons and forks all over and they can stick because now we think there is a metal piece to that." At least she hasn't claimed that the spoons end up being bent after the magnetizing experience. But I know that some gullible minds have been bent.

I've saved the "best" for last. Dr. Steven Hotze is a rabidly anti-gay, anti-vaccine, QAnon-promoting physician who runs the Hotze Health & Wellness Center in Houston. He is also the CEO of the Liberty Center, a political organization that claims the COVID-19 pandemic is part of a "global ritual" to "inject experimental nanobots and chemi-kills into our bodies to alter our DNA using Artificial Intelligence technology to turn us into zombie-like, controlled masses and weapons of war." He makes Dr. Mercola look like a good guy.

According to Hotze, vaccines are an "experimental gene therapy manufactured using cells derived from human babies aborted in the 1970s." Flagrant nonsense. He also claims that in the first month of use, vaccines caused thousands of cases of anaphylactic shock. Codswallop. Data show at most five cases per million injections. Hotze also linked the death of baseball great Hank Aaron to the COVID vaccine, an allegation supported by Robert Kennedy Jr., another anti-vax sage. Aaron was vaccinated, but there is absolutely no reason to link his death to the vaccine. At the age of eighty-six, he was well past the average life expectancy.

Like Mercola, Hotze has also been told by the FDA to immediately correct violations including his promotion of products such as Dr. Hotze's Kids Immune Pak that claim to offer protection against the coronavirus. Previously, this intellectual virtuoso had been challenged for claiming his line of bioidentical hormones prevent cancer, and that birth control pills make women less attractive to men. Wow!

How scientifically educated individuals can go off the rails in this fashion is difficult to explain. Self-delusion, quest for fame, prospects of financial gain, suspect upbringing, a knee-jerk mistrust of "established" science, and the inability to accept being wrong come to mind. Hotze's slogan for vaccines is "Just say no," purloined from Nancy Reagan's anti-drug campaign. I think I will also pilfer it. When it comes to buying into the misinformation and disinformation being spread by the sagacious pundits I've mentioned, just say NO!

SPACE TOURISM

With all the publicity surrounding Richard Branson's and Jeff Bezos's flights into space, you might get the impression that these men are the pioneers of space tourism. Actually, the honor of being the world's first space tourist belongs to American millionaire Dennis Tito, who in 2001 paid $20 million for a ride to the International Space Station (ISS) aboard a Russian Soyuz rocket. An aeronautical engineer who once worked for NASA's Jet Propulsion Laboratory, Tito made his fortune as an investment manager.

Tito had been captivated by Soviet cosmonaut Yuri Gagarin's orbital flight in 1961 and dreamed of following in his footsteps. He described his first sensation of weightlessness as the greatest moment of his life, and his ISS stint as "eight days of euphoria." Seven other space tourists followed, paying the Russians millions of dollars until the program ended in 2009, when the U.S. Space Shuttle Program was retired, leaving the Russian Soyuz craft as the only means of transport to the ISS.

The idea of space tourism was resuscitated on July 12, 2021, when Richard Branson, along with three crewmembers and two pilots, climbed into *SpaceShipTwo*, a winged plane with a single rocket motor that was attached to a specially constructed airplane. At a height of 15 kilometers, the space plane was released, its engine ignited, propelling the vehicle to a height of about 80 kilometers and allowing its occupants to experience a few minutes of weightlessness before landing fourteen minutes later on a runway like a regular airplane.

There is a bit of controversy about whether the flight had actually been a "spaceflight," since where space technically begins is somewhat contentious. Most regulatory agencies accept the Kármán line, defined as 100 kilometers above Earth's mean sea level, as representing the boundary between space and our atmosphere. It is named after aerospace engineer Theodore von Kármán, who was the first to make calculations about where the atmosphere actually peters out. Kármán was born in Hungary, eventually becoming a professor at Aachen University in Germany but because he was Jewish was forced to flee to America with the rise of Naziism. The U.S. Armed Forces and NASA consider 80 kilometers to be the demarcation of space, and by that measure, Branson and his crew are recognized as astronauts.

SpaceShipTwo is a descendant of the X-15 rocket planes of the 1950s and '60s. Carried aloft by a modified B-52 bomber, the X-15 would be detached before the single pilot ignited the rocket engine that burned anhydrous ammonia as fuel using liquid oxygen as the oxidizing agent. This reaction produces nitrogen and water vapor, gases that exit the engine with great velocity. According to Newton's third law that for every reaction there is an equal and opposite reaction, the plane is then propelled in the opposite direction. In 1963, an X-15 reached a height of 108 kilometers and essentially became a spaceship. The pilot experienced a few minutes of weightlessness before hydrogen peroxide thrusters oriented the plane for re-entry into the atmosphere. Aerodynamic flaps of course do not work at that altitude since there is no air.

Richard Branson's flight was similar to that of the X-15, but *SpaceShip Two*'s engine burns hydroxyl-terminated polybutadiene (HTPB), a type of plastic, with the necessary oxygen being supplied by nitrous oxide, which at a high temperature decomposes to yield oxygen and nitrogen. Back in 1914, rocket pioneer Robert Goddard suggested the use of nitrous oxide which was first made by Joseph Priestley in 1772 by the reaction between moist iron filings and nitric oxide. The latter was produced by dropping pieces of iron into nitric acid. Today, nitrous oxide is made by heating ammonium nitrate.

Priestley did not experiment further with what he called "nitrous air diminished," leaving the next step in the development of the gas to the brilliant chemist Humphry Davy. It was he who coined the term "laughing gas" upon noting its mirth-producing effects and also raised the possibility of its use as a painkiller. Laughing gas parties among the British upper class became quickly popular, but the painkilling effect was not capitalized upon until American dentist Horace Wells introduced nitrous oxide to dentistry in 1844. It is still used today to relax patients before undergoing dental procedures.

In the case of Jeff Bezos's flight, a totally different technology was in play. This time, there was no controversy about Jeff and his three mates having earned astronaut wings. The New Shepard rocket, named after America's first astronaut, boosted the crew capsule to an altitude of 107 kilometers, clearly passing the Kármán line. The launch was timed for the fifty-second anniversary of NASA's landing on the moon, and interestingly, the New Shepard rocket has similarities to the *Saturn V* that propelled Armstrong, Aldrin, and Collins towards the moon in that it uses liquid hydrogen as fuel and liquid oxygen as the oxidizer. The total flight took just about eleven minutes, with the capsule making a soft landing in the desert with the use of parachutes and retro rockets. Impressively, the booster also made a successful landing after having exhausted its fuel, ready to be used again.

The stage is now set for future space tourists, with Elon Musk's SpaceX rocket capable of orbital flight set to join the race. So, all aboard! Well, maybe not all. Deep pockets are needed for a ticket: $250,000 for a flight aboard *SpaceShipTwo*, and several million if you want to ride Bezos's Blue Origin rockets. Skyrocketing costs, one might say.

THE FATHER OF MODERN MEDICINE

Sir William Osler once expressed the hope that he would be recognized on his tombstone as the man "who brought medical students into the wards for bedside teaching." His wish never came true for the simple reason that he has no tombstone. Osler's ashes rest in an urn in the Osler Library of Medicine at McGill University, perhaps the most appropriate final resting place for one of McGill's most famous graduates.

Dr. Osler, MD 1872, returned to McGill as a professor in 1874 and introduced the idea of the "journal club," with medical students getting together to discuss the latest research papers. This is an integral part of medical education today. Ten years later he moved on to the University of Pennsylvania as chair of clinical medicine, and in 1889 became the first chief of staff at the new Johns Hopkins Hospital. Osler finished his career at Oxford University before succumbing to the Spanish flu during the 1919 epidemic at the age of seventy.

Described as the Father of Modern Medicine, Osler is revered for his contributions to medical education. His text *The Principles and Practice of Medicine* has educated legions of doctors around the world. Osler maintained that medical students spent too much time listening to theoretical lectures and insisted that his students get to patients' bedsides early in their training. He established the practice of clinical rotations for third- and fourth-year medical students and introduced the idea of residency, whereby newly graduated physicians start their careers in hospitals under a pyramid system, learning from more senior doctors and teaching junior ones.

Under the pseudonym Egerton Yorrick Davis, Osler also wrote humorous articles, one of which reported the supposed phenomenon of "penis captivus," in which it becomes impossible to withdraw the penis after sexual activity. Probably annoyed by the ease with which nonsense was published, he wanted to make a point by submitting his article to the *Philadelphia Medical News*, which uncritically published it. In many ways, Osler's impact on medicine is indisputable, but unfortunately, there are skeletons in the Osler closet.

The first blemish on a formidable career can be traced to a speech in Baltimore that sparked global outrage with his remarks about the "comparative uselessness of men above forty years of age." Osler made reference to an 1882 novel by Anthony Trollope called *The Fixed Period* in which people are dispensed with at the age of sixty-eight to allow youthful talent to take over. It was at this point that he made an off-the-cuff joke about chloroforming people around sixty because they had outlived their usefulness. When newspapers reported the comments out of context, the public became enraged and the term "Oslerize" was coined to describe the extermination of the aged. The doctor became the target of what today we would call "cancel culture" with a deluge of hate mail that even included bottles of chloroform that he was urged to use on himself.

Any suggestion that Osler seriously considered euthanasia for the elderly is nonsense, but he did advocate for early retirement, claiming the world was worse off for allowing so many seniors to stay in positions of power and influence. Indeed, he was of the opinion that "real work of life is done before the fortieth year." One suspects he would have looked askew at the 2020 American presidential race.

Far more serious than accusations of ageism are some of Osler's comments that cannot be described otherwise than racist. "The question with us is what are we to do when the yellow and brown men begin to swarm over," he wrote. "I hate Latin Americans," he is reputed to have said, and in a prank essay under his pseudonym he described some Indigenous obstetric practices in a disparaging fashion, saying that "every primitive tribe retains some vile animal habit not yet eliminated in the upward march of race." On the other hand, Osler advocated for women in medicine, although he also made remarks questioning their competence. He also wrote that "the physician is sent to the sick, and knowing in his calling neither Jew nor gentile, bond or free, perhaps he alone rises superior to those differences which separate and make us dwell apart." Hardly a racist remark.

Osler's views on medicine are eminently quotable. "It is much more important to know what sort of patient has a disease than what sort of disease a patient has." "The good physician treats the disease; the great physician treats the patient who has the disease." "Medicine is a science of uncertainty and an art of probability." Then to students, "I have a confession to make; half of what we have taught you is in error, and furthermore, we cannot tell you which half it is." Bang on!

Where does this leave us with the legacy of Sir William Osler? He was physician with wit and insight who revolutionized the teaching of medicine, but like most of us, made some comments he likely wished he had never made. Dr. Osler is deserving of his place on the medical pedestal, but it should be clear that the pedestal has some cracks.

JAMES BOND AND THE PUFFER FISH

The FBI set up a perfect sting operation. As soon as Edward Bachner picked up the package from his Chicago post office box that day in 2008, he was arrested and charged with the intent of using a biological agent as a weapon. An astute worker at the chemical supply company from which Bachner had ordered the unusually large dose of tetrodotoxin became suspicious and alerted the FBI. This highly toxic chemical is produced by a number of animals, such as the blue-ringed octopus, the rough-skinned newt, some "poison dart" frogs, and, most notably, the puffer fish. It is usually supplied for neurological research in tiny amounts, but Bachner had purchased enough to poison dozens of people.

A quick investigation revealed that to place the order Bachner had misrepresented himself as a doctor at a fictitious lab, that he had ordered tetrodotoxin before, and had taken out a $20 million life insurance policy on his wife. After the arrest, agents searched Bachner's home and found a variety of weapons as well as a book on effective doses for poisoning people.

Tetrodotoxin is an extremely potent, heat-stable neurological toxin that functions by blocking the passage of sodium ions through cell membranes. These ions play a critical role in the transmission of nerve impulses, and interfering with their activity results in signals from nerves not being transmitted to muscles. Should tetrodotoxin enter the human bloodstream, it causes almost immediate numbness of the lips and tongue, progressing quickly to overall muscle paralysis. Death comes from asphyxiation as muscles needed for lung function are paralyzed.

The puffer fish gets its name from an ability to inflate its body like a balloon when threatened, exposing sharp, tetrodotoxin-loaded spikes on its skin. Should a predator not be deterred by the spikes and attempt to make a meal of the fish, it likely will be its last one. Humans are also puffer predators, since the flesh of the fish is regarded as an exotic delicacy, especially in Japan. The preparation of fugu, as the puffer fish dish is known, has to be done very carefully to ensure that all the toxin-containing parts are scrupulously removed. Chefs are carefully trained and have to prove their competence by eating a fugu dish they have themselves prepared before being allowed to serve diners in a restaurant.

Thanks to the extensive training, restaurant accidents now are very infrequent but do occasionally occur. Witness the misadventure of Homer Simpson, star of the animated sitcom *The Simpsons*. Homer becomes a victim of fugu poisoning when he is served fish improperly prepared by an apprentice who is pressed into service while the master chef engages in sexual exploits behind the restaurant. When told he has twenty-two hours to live, Homer goes through the five stages of grief, but luckily survives, as some people do. James Bond, Ian Fleming's famous secret agent, also makes it through a bout with tetrodotoxin but needs some help from science. In Fleming's novel *From Russia, With Love*, Bond is in a fight with a Russian agent who is wearing a boot that can flash out a small blade coated with tetrodotoxin. A quick kick to Bond's shin and the novel ends with the reader left hanging about whether 007 survives. Only in the next novel, *Dr. No*, do we learn that Bond received immediate artificial respiration, which is critical if someone is to survive tetrodotoxin. Then a physician diagnosed him with curare poisoning and administered appropriate treatment. Here Fleming's science can be called into question.

Curare is a neurotoxin that is extracted from a vine that grows in the jungles of South America and has a history of being used as a poison on arrows. Like tetrodotoxin, it is a neurotoxin but works by a different mechanism. Tubocurarine, the active ingredient, produces paralysis by blocking the action of the neurotransmitter acetylcholine. This can be reversed with drugs such as physostigmine that inactivate acetylcholinesterase, the enzyme that normally degrades acetylcholine. As a result, levels of acetylcholine increase sufficiently to displace curare from receptors. However, acetylcholinesterase inhibitors would not work in the case of tetrodotoxin poisoning since acetylcholine is not involved. There is actually no antidote to tetrodotoxin poisoning. Apparently, Bond was not fazed by his close encounter with death because in *You Only Live Twice* he happily dines on fugu.

And what happened to Edward Bachner? He was sentenced to almost eight years for illegally acquiring a dangerous material but was not charged with attempted murder. The defense argued that he got involved in a bizarre fantasy game in which he aimed to show how murder with a poisonous chemical could be carried out, but never intended to actually follow through. His wife stood by him, saying that she never believed her husband ever had any intent to harm her.

Tetrodotoxin makes another appearance in the Bond film *Octopussy*, in which Bond's love interest is named after the poisonous blue-ringed octopus she keeps as a pet. As may be expected, the octopus does its thing and ends up wrapped around the face of a villain. Death by octopus.

GUTTA-PERCHA, WALKING STICKS, AND HICKORY GOLFERS

In the 1800s a fashionable European gentleman would wear a top hat and carry a walking stick that served both as a decorative dress accessory and as a self-defense item against street crime. Although not quite as popular as in Europe, walking sticks were part of the American scene as well, serving as a status symbol. In one particular case, though, a walking stick was not used as protection against crime, but rather to commit one.

That walking stick, or at least a broken part of it, can be seen displayed at the Old State House in Boston. There are two interesting features of this relic, one scientifically noteworthy and one historically disturbing. The stick is made of gutta-percha, the hardened latex of the *Palaquium gutta* tree, originally native to Malaysia. This is a natural thermoplastic substance, meaning it can be softened with heat and shaped into a form that is retained on cooling. Gutta-percha was introduced to Europe in 1842 by Dr. William Montgomerie, a surgeon serving with the British army in the East Indies who had originally come across the substance in Singapore, where it was being used to make handles for machetes. He thought the substance would be useful to produce handles for medical devices as well as splints for fractures.

Victorian society quickly took to gutta-percha. Chess pieces, mirror cases, and jewelry were fabricated with it, and dentists found it useful for filling cavities. But perhaps the biggest impact was on the game of golf. At the time, golf balls were made of feather-stuffed leather, were expensive, and not exactly aerodynamic. Balls fashioned out of gutta-percha were cheaper and flew farther. When they were dinged up, these "gutties" could be repaired by softening in boiled water and then reshaping in a hand press. The ball's popularity increased when it was discovered that grooves cut into the surface allowed for a longer flight. Gutties were the ball of choice until about 1900, when they were replaced by the Haskell ball made of a solid core of rubber wrapped tightly with rubber threads.

Interestingly, rubber, which is also an exudate of a tree, and gutta-percha have almost identical molecular structures. They are both polymers of a simple molecule, isoprene, so can be termed as polyisoprenes, but different "kinks" in the long molecules, referred to as "cis" or "trans," allow for different properties. While gutta-percha is thermoplastic, rubber is thermosetting, meaning that once formed into a shape it cannot be reshaped with heat. The rubber used in the Haskell ball was vulcanized, a process introduced by Charles Goodyear, who discovered that treating natural rubber with sulfur allowed it to be made into a very hard material. It turns out that the sulfur atoms cross-link the cis polyisoprene units to form a tough latex.

Michael Faraday, the brilliant English scientist who carried out numerous experiments with electricity, found that gutta-percha was an excellent insulator, a property that allowed it to be put to use as a coating for the newfangled telegraph cables. In a monumental engineering undertaking between 1854 and 1858, the first transatlantic telegraph cable, insulated with gutta-percha, was laid down. Unfortunately, it quickly failed. But by 1865, improvements in technology resulted in a properly functioning gutta-percha insulated telegraph cable that allowed messages to be sent between the continents in a few minutes. Prior to this, communication was via ships and could take weeks. Gutta-percha proved to be a huge triumph and served well until eventually replaced by polyethylene insulation.

Now on to that gutta-percha walking stick exhibited in Boston's Old State House. It was used in the famous "caning of Charles Sumner" case, a significant blemish on American history. In 1856, Democrat Preston Brooks brutally attacked Republican Charles Sumner with his walking stick on the floor of the U.S. Senate in the Capitol. Sumner, a dedicated abolitionist, had made a strong speech against slavery, a practice that Brooks favored. The attack was so violent that Brooks's gutta-percha cane broke into pieces, some of which were recovered from the Senate floor and cut into rings that Southern lawmakers wore on neck chains to show their solidarity with Brooks, who boasted that people begged for pieces of his cane as sacred relics. The caning was followed by demonstrations in northern cities to support Sumner, and in the South to support Brooks. That gutta-percha walking stick is a stark reminder of the division in the U.S. that led to the Civil War, a division that still exists to this day, albeit based on different ideologies than back then.

Gutta-percha is mostly a relic of the past, having been displaced by a variety of high-performance polymers. Except in one application! It is still used in dentistry, although not to fill cavities as in the early days. Gutta-percha is the best material with which to fill root canals after diseased tissue has been removed. And there is still a market for gutta-percha golf balls! Hickory golfers, a surprisingly large global community, purchase them to play rounds as they would have been played in the 1800s. They use only clubs with shafts of hickory wood instead of metal, and "gutties" made by a process like in those nostalgic days of yore when a drive of 160 yards was the best one could hope for.

JOHN DEE AND 007

Amazing staging! Trygaeus climbs onto a giant mechanical beetle and flies up to the palace of the gods. That scene from Aristophanes's play *Peace* would be spectacular on the Broadway stage today, but what is truly amazing is that it was orchestrated in 1547 at Trinity College, Cambridge! The brains behind the spectacle was John Dee, a young faculty member who instantly developed a reputation as a sorcerer because the audience could not believe that such a spectacular effect could be produced by normal means.

Dee would go on to forge a career as mathematician, astronomer, navigational expert, cartographer, book collector, and alchemist. He would certainly qualify as a scientist if the description ended there. But it doesn't. While the Cambridge beetle had nothing to do with sorcery, Dee later would go on to live up to the reputation it had fostered by dabbling in astrology, exploring contacts with the spirit world, and engaging in fortune-telling. A curious mélange of science and the occult!

In 1558, Dee cemented his status as a seer by advising young Princess Elizabeth not to despair because, "as the gods have indicated to me, you shall become Queen in another four months." Indeed, exactly four months later, Mary Tudor died, allowing Elizabeth to ascend to the throne. Out of gratitude, the queen appointed Dee as her personal astrologer and advisor.

Before long, John Dee proved to be so useful that he was given the task of gathering intelligence about foreign rulers and reporting directly to the queen. A secret agent as it were! These reports were not signed with his name, but rather with a symbol of two circles flanked with a horizontal and a vertical line that can be interpreted as the number seven. Supposedly the circles represent eyes, meaning the report was only for Her Majesty's eyes. The number seven was there because it was the alchemists' lucky number. And there we have the first secret agent, code name 007! Is this where Ian Fleming got the idea for 007? Was John Dee the inspiration for James Bond? We will never know because Fleming is no longer with us. An alternate theory is that Fleming's research into spy activities revealed that one of the great British successes during World War I was the cracking of a German code that the British referred to as 0070. The author just shortened this to 007. I prefer the association with Dee because he was into chemistry!

This is documented in a famous nineteenth-century painting by Henry Gillard Glindoni depicting Queen Elizabeth and her courtiers watching Dee performing a chemical experiment. "The queen's conjurer" is clearly seen pouring some substance from a vial into a flaming brazier. Looks like a demonstration we commonly carry out in chemistry lectures, sprinkling a little lycopodium powder into a flame. Dee documents other chemical experiments in his writings, including the making of silver chloride. Although not completely clear, it seems he reacted silver with nitric acid to form silver nitrate, which then yields silver chloride on reaction with salt, sodium chloride.

Dee's interest in chemical matters is further demonstrated by his association with the infamous occultist, self-declared spirit medium, and alchemist Edward Kelley. Having once been convicted of forgery, Kelley's ears were cropped as punishment, and we see him in Glindoni's painting with a hat that covers the disfigurement. Kelley had sought Dee out in 1582, offering his help as a medium upon hearing of Dee's efforts to foretell the future by gazing into a mirror. Indeed, Dee was into "scrying" with a mirror made of obsidian, a volcanic rock. Kelley claimed that he had the ability to contact angels who would help him interpret the visions Dee saw in the mirror.

The mirror that John Dee is supposed to have used is on display in the British Museum along with his crystal ball. Such obsidian mirrors were introduced into Europe by Spanish explorers who had found Mexican natives using them for divination. The required shine was imparted to the rock by rubbing with, get this, bat droppings! Since bats only partially digest insects they eat, residues of the bugs' exoskeletons show up in the feces, making this a functional abrasive to polish the volcanic rock. Also on display in the British Museum is a clay tablet with all sorts of occult symbols that Kelley used in his communications with angels. He would interpret the messages for Dee, including the famous one about the need to share all earthly possessions, including wives. And yes, that meant Dee and Kelley came to engage in wife swapping! One of Dee's children may actually have been fathered by Kelley.

Dee and Kelley traveled through Europe with Dee telling fortunes and demonstrating scientific phenomena, while Kelley attempted to mutate metals into gold with a magic powder he claimed to have discovered. In Bohemia Dee was even imprisoned for a while after he failed to produce the metal as promised. He eventually fell out of favor when Elizabeth was succeeded by James I who abhorred divination and anything to do with the occult. The enigmatic polymath who had once been the toast of the royal court died in poverty.

A recent x-ray analysis of the Glindoni painting shows that originally Dee was surrounded by a circle of skulls, which the painter may have intended to portray that while Dee pursued science, he also had a foot in the occult. Perhaps the patron who commissioned the painting did not like this association and asked the artist to modify the work. He painted over the skulls, and perhaps piqued, retained the occult connection by replacing a globe in the original with Kelley. I just wish he would have sneaked the 007 in there somewhere.

MARASCHINO CHERRIES

A banana split with a cherry on top! That was the first dessert to which I was ever treated in Canada. Of course, we had ice cream in Hungary, but I had never seen a banana. And I certainly had never seen anything like the dazzling cherry that practically glowed atop the enormous, chocolate-drenched scoops of ice cream. My first encounter with a maraschino cherry! It would not be my last. But it would be the chemistry, not the taste, of this strange little specimen that would eventually attract my attention.

The birthplace of the maraschino cherry that enticingly bobs in many a cocktail, is the lab. Its ancestry, though, traces back to 1905, when black Marasca cherries, preserved in a sugar-sweetened liqueur, were introduced by the Luxardo distillery in Croatia. The spirit, christened "maraschino," was distilled from a fermented mix of cherries, pits, leaves, and stems, hence the name of the cherry bathed in it. Unlike fresh cherries that bruise and spoil easily, maraschinos kept well and traveled well. By the late 1800s, they had crossed the Atlantic, and with their flair for tempting bar patrons soon won the hearts of bartenders.

Importing European maraschinos was expensive, so with typical ingenuity, Americans began to experiment with cheaper versions. Since Marasca cherries were scarce, the Royal Ann variety was enlisted with various methods of preservation other than soaking in alcohol being explored. A solution of sodium metabisulfite was found to work particularly well. There was a problem, though, because the sulfur dioxide released destroyed the fruit's color and flavor! Synthetic coal-tar dyes solved the color problem, and benzaldehyde was added to boost taste. Benzaldehyde is the characteristic flavor of almonds, but also occurs naturally in cherries. When used as an additive, it is generally produced synthetically, but that is irrelevant. Benzaldehyde is benzaldehyde, whether extracted from natural sources or produced in the lab.

These "imitation maraschino cherries" looked and tasted nothing like the original, but they served well as eye candy in drinks and atop sundaes. Actually, they were more than just eye candy. Pumped full of sugar to satisfy American taste buds, these cherries were closer to being a candy than fruit.

There was yet another problem. With time, the cherries tended to turn mushy. In 1911, one unhappy food critic opined that "long imprisonment in a bottle" reduced these chemically manipulated cherries "to a formless, gummy lump" and predicted that this "abomination" would disappear once its "utter unfitness has been manifested."

But the cherries did not disappear, due in large part to the work of Professor Ernest Wiegand at the Oregon Agricultural College, now Oregon State University. In 1925 Wiegand began to tackle the "fitness" issue with his efforts being welcomed in a state where farmers had invested heavily in growing Royal Ann cherries. To this day, the university offers a whole course on the maraschino cherry.

Wiegand worked for six years before his "aha moment," the discovery that calcium chloride added to the metabisulfite bleaching solution cross-linked pectin polymers in the flesh of the cherries, resulting in a firm texture. Mushiness gone! Dr. Wiegand also introduced sodium benzoate and potassium sorbate to prevent microbial contamination, and demonstrated that with a pH maintained below 4.5, any Clostridium botulinum bacteria present are unable to produce highly toxic botulin.

At the time, the most suitable red dye was FD&C No. 4, but its use ran into a problem in 1965, when a study in dogs linked it to adrenal gland and bladder toxicity, albeit at levels much higher than found in maraschino cherries. It was eventually replaced either with Red No. 40 (Allura Red) or Red No. 3 (Erythrosine), both of little concern since their metabolites are excreted in the urine. Issues have been raised about some synthetic dyes possibly affecting children's behavior, but that cherry on a sundae is not going to make any kid bounce off the walls.

By now, you may be thinking that the current version of the maraschino cherry belongs in the "Frankenfood" category, and wondering whether the original Luxardo cherries are available. Yes, they are! During World War II, the distillery was essentially destroyed and many in the Luxardo family were killed. Giorgio Luxardo survived and, taking along some Marasca saplings, relocated to Italy. Soon a newly built distillery was producing Maraschino Originale liqueur as well as the fabled cherries. No artificial colors, flavors or preservatives. No scrimping on sugar, though.

Since cherry pits are part of the liqueur's recipe, people have wondered about the presence of cyanide, given that crushing the pits releases benzaldehyde and hydrogen cyanide. A curiosity here is that both benzaldehyde and hydrogen cyanide smell like bitter almonds. Remember those mystery stories where a detective detects the smell of almonds on a corpse's mouth and concludes that murder has been committed? No need to worry about maraschino liqueur, though; there is hardly a trace of cyanide per serving. And the "fake" maraschinos?

They do smell of almonds, but no cyanide in sight. Just a whiff of benzaldehyde added as flavoring.

A 2005 *New York Times* article poked fun at the nutritional composition of the highly processed maraschinos and denigrated them to being the "culinary equivalent of an embalmed corpse." Come on now, nobody eats maraschinos for their nutritional value. They just provide a little fun, a little kitsch.

But if you want the real deal, the kitsch-free Luxardo cherries, you can expect to pay handsomely, roughly $1.60 per cherry. So, you don't want to plunk these into the kids' Shirley Temple drink. But they will add a whole new dimension to a banana split.

KEEP THAT TEMPERATURE LOW

"When I tasted the meat, I was very much surprised indeed to find it very different, both in taste and flavor, from any I had ever tasted." With that observation, recorded in his 1802 *Essays, Political, Economical and Philosophical*, Sir Benjamin Thompson launched the science of low temperature cooking that a couple of centuries later would evolve into the sous vide technique.

Thompson was an established scientist by the time he turned his attention to cooking, albeit a scientist with an unusual history. Born in Massachusetts, he developed an early interest in science and became a schoolmaster in Rumford, New Hampshire, where he married a wealthy older widow. As the American Revolution was brewing, he sympathized with the British, and after his house was attacked by a mob, Thompson fled for refuge behind the British lines. Here he was welcomed, both because he was able to offer information about the American forces and because he was interested in experimenting with improving gunpowder.

After the revolution, Thompson fled to England where he became royal scientist to the king and then minister of war. He ran into trouble with King George III after supposedly selling naval secrets to the French but was spared punishment when he agreed to be sent as a diplomat to Bavaria and act as a spy. Once there, he severed his British ties and became a scientific consultant to Karl Theodor, Bavaria's ruler. The country was beset with poor people who had no work, and Thompson was given the task of solving the problem. He came up with the idea of workhouses, where the poor could earn some money while serving as a source of low-cost labor. The question of how these workers could be adequately and cheaply fed triggered an interest in food and nutrition.

Thompson concocted an economical soup based on pearl barley, yellow peas, potatoes, and beer that according to the knowledge of the day was nutritious. For his scientific efforts, he was made a count, and when asked to choose a name, he chose "Count Rumford," an homage to the town where he got his start. His soup became a common, inexpensive military ration and was christened "Rumford's Soup." Further kitchen experiments also led to the percolating coffee pot, the double boiler, an improved pressure cooker, a smoke-free fireplace, and a machine for drying potatoes. It was the latter that would lead to Thompson being recognized as the pioneer of low-temperature cooking.

"Desirous to find out whether it would be possible to roast meat in a machine I had contrived for drying potatoes, I put a shoulder of mutton into it, and after attending to the experiment for three hours, and finding it showed no signs of being done, I concluded that the heat was not sufficiently intense and I abandoned my shoulder of mutton to the cook-maids." After putting out the fire under the dryer, the maids figured the meat could be stored in it overnight as well as anywhere else. In the morning, they were surprised to find the meat cooked and "not merely eatable, but perfectly done, and most singularly well-tasted." Thompson concurred, and concluded that "the gentle heat over a long time had loosened the cohesion of its fibers, and concocted its juices, without driving off their fine and more volatile parts, and without rendering rancid and empyreumatic its oils."

Now there is a neat word for "having the taste or smell of slightly burned animal or vegetable substances."

Thompson went on to say he "had long suspected that it could hardly be possible that precisely the temperature of boiling water should be the best adapted for cooking all sorts of food." At the time, cooking in vigorously boiling water, or roasting at high heat, were the standard methods in the kitchen. Thompson began to challenge this wisdom by demonstrating that gently boiling water would serve just as well since the more intense boiling did not increase the temperature. He also pointed out that the extra heat needed for vigorous boiling wasted fuel!

Low-temperature cooking flew under the radar until French chef Georges Pralus discovered in 1974 that foie gras cooked at a low temperature in a vacuum-sealed plastic bag immersed in water kept its original appearance, retained its fat content, and had a better texture. Other chefs began to experiment with sous vide cooking, which means "under vacuum," and discovered that the vacuum was not essential and the technique would work as well by squeezing out the air from a plastic bag. As devices allowing for precise control of temperature in a pot of water became commercially available, home cooks began to experiment and found the results to be very satisfactory, except for one troubling point.

Meat, while tender and flavorful, did not develop the usual brown crust. That is because the Maillard reaction (named after the French chemist Louis Camille Maillard, who discovered it in 1912) cannot occur at temperatures below about 135°Celsius. In this reaction, amino acids react with carbohydrates in the food to produce an array of brown and tasty compounds, as evidenced by the crust that forms on bread and the grill lines on steak. Chefs counter this problem by quickly searing meat after sous vide cooking.

The absence of the Maillard reaction has some positive features, given that some of the products, such as acrylamide, are purported carcinogens. A quick sear, though, is safe enough. Another question that comes up concerns the leaching of chemicals from the plastic bags, but the bags specifically made for sous vide are generally polyethylene, which does not lead to any leaching of estrogenic compounds, a worry with some plastics such as polycarbonate or polyvinyl chloride.

Together with Sir Joseph Banks, Count Rumford founded the Royal Institution of Great Britain and championed science lectures for the public. On his death, he bequeathed funds to establish the Rumford Professorship at Harvard University. In the nineteenth century, this chair was held by Eben Horsford, who developed the first commercial baking powder. In honor of Benjamin Thompson, he named it Rumford Baking Powder. It is still sold today.

THE YELLOW SCHOOL BUS

You do not usually find a mix of engineers, transportation officials, paint producers, and car manufacturers at an educational conference. But there they were, hobnobbing at Columbia University. The year was 1939 and they had been invited by education professor Frank W. Cyr who was concerned about the haphazard means by which students were getting to school and aimed to establish some sort of national standards for school buses. Over seven days, standards for length of bus, ceiling height, and aisle width were ironed out. When it came to a discussion of color, sample strips ranging from yellow to red were hung on the wall, and all agreed that yellow would be the most visible, especially in the dark. It also turns out that yellow is the color most easily seen from the corner of the eye, so drivers would readily become aware of a school bus approaching on either side.

Cyr and his invited guests were not the first to note the high visibility of yellow vehicles. Even before the advent of the automobile, horse-drawn cabs in London and Paris were colored yellow. As early as 1798, theatergoers in Paris flocked to the hit musical *Le cabriolet jaune* (*The Yellow Cab*), that featured jealousy, mistaken identity, and two yellow cabs. (Our term "cab" actually comes from "cabriolet.")

How did they paint the cabs yellow in those days? Possibly with one of the pigments known since antiquity. Yellow ochre, for example, is a naturally occurring form of iron oxide, yellow arsenic sulfide is found in the mineral orpiment, and "Indian Yellow" can be isolated from the urine of cows fed mango leaves. Even synthetic pigments such as lead antimonate, prepared by heating lead oxide with antimony oxide, were known. But these gave way to a brilliant "Chrome Yellow," thanks to a discovery by French chemist Nicolas-Louis Vauquelin, who in 1797 isolated a new element from the mineral Siberian red lead. In combination with other substances it produced a range of colors, so he named it "chromium" from the Greek for "color."

Chromium's reaction with oxygen and lead yielded the yellow pigment lead chromate that when ground into a powder and mixed with linseed oil formed Chrome Yellow. Artists took immediate note of this oil paint, as did cab companies in Paris and London. Their cabs were now easily distinguished from other carriages. When automobiles appeared on the scene, the tradition continued. In 1907, John Hertz (yes, that John Hertz) launched a fleet of yellow cabs in Chicago, and two years later yellow cabs began to cruise the streets of New York and have never left. By 1939, when it came to making a decision about the color of school buses, there was no question about what paint would be used. "National School Bus Chrome" would become the standard.

Artists were especially fond of Chrome Yellow, prizing it for its stunning brilliance. Van Gogh's famous sunflowers dazzled with the pigment! Unfortunately, they don't dazzle quite as much today. In 2011, researchers found that that ultraviolet light triggers the release of electrons from soil contaminants in the paint that then reduce chromium ions from a +6 to +3 oxidation state. Chromium compounds that include ions with a +3 charge are dark brown. To limit fading, Van Gogh's masterpieces are now displayed with limited exposure to light.

It is not only the sunflowers series that are a victim of aging, the yellows in Edvard Munch's classic *The Scream* are also losing their luster, although for a different reason. Here the problem is humidity.

Munch was already aware of the problems with Chrome Yellow and turned to a novel pigment that had recently been introduced. German apothecary Friedrich Stromeyer was making up some calamine lotion when he noted that the shipment of zinc carbonate he had ordered was not as white as usual. Experiments to trace the problem led to an impurity from which he isolated a substance that turned out to be a new element. He named it "cadmium," from the Latin "cadmia," the ancient name for zinc carbonate. When cadmium was combined with sulfur, it produced Cadmium Yellow, a substance Stromeyer noted "promises to be useful in painting." Indeed, it was. But not quite as useful as it first seemed. Like Chrome Yellow, it also faded with time.

The chemistry here has now also been elucidated by comparing samples from unopened tubes of Munch's paints to the yellow in *The Scream*. It turns out that on exposure to humidity, cadmium sulfide turns into cadmium sulfate, which is white. I am troubled by that because I have a special fondness for this painting, having used it in many a lecture because to me it represents the anxiety produced by alarmists who claim that our health is being threatened by the likes of vaccines, GMOs, food additives, and 5G networks.

After it was stolen in 2004 and recovered in 2006, *The Scream* was put in climate-controlled storage but in 2022 is back on display in Oslo's new National Museum in a room where humidity is carefully regulated.

As far as school buses go, they have not been painted with Chrome Yellow since the 1970s when lead-based paints were taken off the market. The paint never presented a risk to students, but occupational exposure for workers who produced it was a concern. The paint used today is School Bus Glossy Yellow, which has a number of proprietary formulations. Some versions are made by mixing red and green paints, others use the synthetic organic compound benzidine yellow.

Hopefully, those yellow school buses will soon be loaded with students again on their way to being educated, maybe even about why sunflowers are called sunflowers and why chemists think of *The Scream* when they see ads for "Chemical-free" sunscreens.

NO, IT DOESN'T SWITCH MY STEM CELLS ON

I thought I was going to write yet another depressing column about a properly trained medical doctor going astray and trading his soul for profits by promoting questionable dietary supplements. Sometimes, though, you start a journey with a frown and end up with a smile. But let's start at the beginning.

I'm used to getting emails that attempt to hook me into watching interminable videos about some "cutting edge breakthrough" that will rejuvenate me or at least "turbocharge my brain." Usually it is some sort of "natural" plant derivative discovered by a maverick scientist who has managed to sidestep the efforts of Big Pharma to sweep this miracle under the carpet. I'm told I can put spring back into my step and restore the twinkle in my eye as my cells are "flooded with a tsunami of lifegiving nutrients that have never been made available to the public before." But I can have them now, while supplies last, for half price, if I order in the next ten minutes.

These days, when another one of these travesties against science beckons with a promise to make my "body fire on all cylinders" with some superfood, I just hit "delete." However, just as I was ready to relegate a message from the *Alternative Daily* to trash, my eyes got stuck on a sentence. All I had to do to protect my brain was to "avoid certain foods that contain a dangerous enzyme called diacetyl." What is wrong with that? Diacetyl is not an enzyme; it is a small, simple molecule that occurs naturally in butter but can also be produced synthetically to impart a buttery taste to popcorn. What's dangerous about it? In an occupational setting, as in workers who inhale diacetyl while producing artificial butter flavoring, it has been linked with bronchiolitis obliterans, a rare lung disease, popularly referred to as "popcorn lung." It has never been associated with any problem when consumed.

The *Alternative Daily* attributed the diacetyl-brain connection to "brain health specialist and NASA scientist Dr. Sam Walters," who claims that "diacetyl often passes through the blood brain barrier and can lead to deposits in the brain." There is no evidence for this. I was then directed to a video by Walters offering the usual tropes about MSG, aspartame, sucralose, and aluminum being "Mental Menaces," and the need to counter their assault with a "one-of-a-kind, multi-spectrum solution" for age-related memory loss that this brilliant scientist had formulated.

Calling diacetyl an enzyme and shilling for a supplement suggested that I had the makings of a "doctor gone astray" story, but it wasn't long before I was stopped in my tracks. It turns out that Walters isn't a medical doctor, he is a naturopath! So, he hadn't really strayed from his profession at all, since dietary supplements and vilification of anything that isn't "natural" are part and parcel of his trade. Furthermore, being a naturopath hardly qualifies one to be a "brain specialist," and I was unable to find any evidence that he is a "NASA scientist." I did learn that he is into homeopathy, spinal manipulation, hyperbaric oxygen therapy, and, of course, herbal medicine.

The Youthful Brain supplement promoted in the video lists as ingredients "1,000 mcg Vitamin B12 dosed at 16,667% Daily Value for maximum mental energy, 280 mg total of Bacopa monnieri, phospha-tidylserine, *Ginkgo biloba* leaf extract and huperzine A." Vitamin B12 does not give you mental energy, and at 16,667 percent of daily value (ridiculously given to five significant figures) you end up peeing most of it out. Bacopa monnieri extract is included because it is supposedly favored by "a secret society of Tibetan monks with steel-trap memories who rarely fall victim to sickness." Right.

Why *Ginkgo biloba*? Because we learn that ginkgo leaves are the favorite leaf of the elephant, and elephants have spectacular memories. What more evidence can we want? Huperzine, extracted from a type of moss, increases levels of the neurotransmitter acetylcholine, which is indeed important for proper functioning of the nervous system. Phosphatidylserine, found in the membranes of brain cells, can be produced from soy lecithin and is "the memory molecule of youth," according to Walters. A few studies have indeed demonstrated some beneficial effects for huperzine and phosphatidylserine, but not at the low doses found in Youthful Brain!

Walters also pushes Stem Cell Renew, with its content of wild blueberry fruit powder, organic goji berry, grape extract, and ginkgo. The rationale? Walters claims that the inhabitants of Bapan, China, known as the Longevity Village, owe their long lives to taking such a herbal mix every day. There is absolutely no evidence they do this. Furthermore, renewing stem cells is nonsense.

Bapan villagers really do appear to have unusual longevity. It is likely due to being active, having a strong social fabric, consuming a mostly plant-based diet, and being happy with their lives. They smile a lot! And believe it or not, that is predictive of a longer life. Want some scientific evidence? Researchers examined baseball cards released during the 1952 season and rated the photos as "no smile," "partial smile," or "full smile." The average lifespan for the "no smilers" was seventy-three years, for the "partial smilers" was seventy-five years, and for the "full smilers," eighty years! The results were deemed to be statistically significant. How can this be rationalized? Perhaps people who smile readily are less likely to become stressed as they are buffeted about by the waves of life. Stress is definitely linked with reduced longevity!

I guess instead of frowning with annoyance when some wonder product claims to "flip my stem cell switch on," I should just smile at the folly of such a promise.

THE TRUTH IS OUT THERE

"The truth is out there." That was the catchphrase for *The X-Files*, the popular sci-fi series that debuted in 1993. The program certainly caught my attention because at that time I had already been focusing on the public's understanding of science, pushing the importance of critical thinking and the need for basing conclusions on observation.

In presentations to students and the public I often used the example of unidentified flying objects, or UFOs, to illustrate how easily one can arrive at a wrong conclusion based on an observation. Many observers have concluded that they have seen something unworldly, perhaps even a proverbial flying saucer, when what they actually saw were unusual cloud formations, missile launches, camera artifacts, aircraft, light reflections, balloons, satellites, birds, or planets, Venus in particular. Some have even been the victim of clever hoaxers.

I was particularly attracted to *The X-Files* because it featured a pair of FBI agents who were assigned to investigate cases that seemed to be outside the realm of science. Mulder believed in assorted paranormal phenomena as well as in alien visitations, while Scully was a skeptic, a scientifically trained doctor, who usually came up with alternate rational conclusions to Mulder's mystical ones. That was right up my alley.

My interest in the "observation-conclusion" process can be traced to my taking up magic as a hobby. After all, the goal of a magic performance is to steer the audience to the wrong conclusion about what they are observing. When you show an empty hat from which you proceed to pull a rabbit, you expect the audience to conclude that the rabbit magically materialized out of nothing, totally contrary to the laws of nature. And yes, I used to perform this effect with a live rabbit, until Ether outgrew the hat. Now I use a synthetic rabbit. Needless to say, it is all done by scientifically explicable means.

Curiosity about UFOs, a natural spin-off from working with illusions, was triggered by a talk from physicist Stanton Friedman sometime in the 1980s. Trained as a traditional physicist, Friedman forged a new career as a "ufologist" after investigating the supposed landing of an alien craft near Roswell, New Mexico, in 1947. He became convinced that intelligently controlled extraterrestrial spacecraft were cruising our skies and that the government was involved in a huge cover-up. Friedman sounded very credible and showed a press release from the Roswell Army base about a "flying disk" having been captured as well as a retraction the next day with an explanation that a mistake had been made. What had actually been recovered were the remains of a weather balloon! I was intrigued, and down the rabbit hole I went.

It seems that the first mention of "flying saucers" was by Kenneth Arnold, who in 1947 was piloting his small plane near Mount Ranier in the state of Washington and reported seeing a group of objects traveling at high speed, moving "like saucers skipping on water." A newspaper account mistakenly reported that the objects were saucer-shaped and the mythology of the flying saucer was born. Around the same time, farmer Mac Brazel came across debris near Roswell that he took to the local sheriff, who later reported that Brazel "whispered kinda confidential-like" that he might have found a "flying disk." Whether Brazel was aware of Arnold's sighting is not clear. The sheriff in turn contacted Major Jesse Marcel, an intelligence officer at the local army base, and an investigation began, resulting in the press release and subsequent retraction. The whole affair was a minor news story at the time and soon faded.

Friedman's investigation began in 1978 when he encountered Major Marcel and heard about how he had actually handled the remains of a UFO. The physicist then began to interview people who were in any way involved with the supposed crash. This was almost thirty years after the event, and as we know, memories are rather malleable! In any case, Friedman concluded that an alien craft had actually landed and, for some reason, the epic event was covered up.

I got hooked, collected books, articles, listened to expert interviews, and even attended a UFO conference. It turns out that there was indeed a cover-up, just not the kind promoted by ufologists. In 1947, the U.S. government had launched "Project Mogul" aimed at detecting sound waves generated by Soviet atomic bomb tests using high altitude balloons equipped with microphones. It was one of these that landed near Roswell, and since the project was top-secret, the government preferred that the crash be portrayed to the public as a "flying disk." When it was clear that the remnants being exhibited did not look like pieces from a spacecraft, the story was changed to a weather balloon. Of course, there are still devoted ufologists who believe that the cover-up was a cover-up. It wasn't.

Since 1947, thousands of UFO sightings have been reported, mostly in North America, apparently the aliens' preferred continent. The vast majority are explained, but many remain "unidentified." Interest geared up in 2021 thanks to a report by the Pentagon that examined videos taken by aircraft from the carrier *Nimitz* in 2004 that appear to show "Unidentified Aerial Phenomena." Several pilots have claimed to have seen a craft that looks like a giant Tic Tac with no obvious means of propulsion. Experts have already offered opinions about the sightings being camera artifacts, glare from engines, or, yes, balloons.

As for me, I would be thrilled to have evidence of contact with aliens. It would make science even more fascinating. While I think it extremely unlikely that life evolved only on our insignificant little planet in an insignificant galaxy, I think it is just as unlikely that we are being visited by extraterrestrials. The distances are just too great. The truth is indeed out there. Way, way out there. Just like some theories about UFOs.

DENTAL IMPLANTS

The poverty-stricken people who queued up outside Dr. John Hunter's home in London were not there to be treated for any disease. They were there to have their teeth yanked out. Without the benefit of an anesthetic!

Word had gotten around that the doctor would pay handsomely for extracted teeth that he would then attempt to implant into the gums of the gap-toothed rich. Hunter, an early advocate of applying the scientific method to medicine, had hatched this idea after carrying out an unusual experiment he described in his 1778 publication, *Treatise on the Natural History and Diseases of the Human Teeth*. After making an incision in the comb of a rooster, he implanted a human tooth, noting that "the external surface of the tooth adhered everywhere to the comb by vessels, similar to the union of a tooth with the gum and sockets."

Roosters were undoubtedly not a fan of Hunter's work since they were also the subjects of his testicular transplant experiments. He had noted that the combs of castrated roosters drooped, but became erect again after reinserting a testicle into their abdominal cavity. Although not recognized at the time as such, this was an early demonstration of hormonal activity!

Hunter's rooster head with the implanted tooth has been preserved and is on display in London's Hunterian Museum along with hundreds of other anatomical specimens, including a rather controversial skeleton, that of seven-foot, seven-inch Charles Byrne, whose body Dr. Hunter had arranged to steal on the way to his funeral. Neither was Hunter averse to dealing with grave robbers to supply corpses for his anatomical studies.

Dr. Hunter's transplanted teeth never adhered properly to bone and did not fare well. Indeed, to this day, no tooth transplants have been successfully performed. However, artificial crowns anchored into bone with titanium screws have revolutionized dentistry, at least for those able to afford such implants.

The idea of replacing lost teeth with artificial ones is by all means not new. The ancient Egyptians and Phoenicians managed to bind human teeth, or ones carved from ivory, to existing teeth with gold wire, and the Mayans made attempts to replace missing teeth in the lower jaw with pieces of sea shells. But only in the latter years of the twentieth century would dental implants become a practical reality thanks to the discovery that metallic titanium fuses with bone without any sign of rejection.

Drs. Bothe in Great Britain and Leventhal in the U.S. had been experimenting with using metal implants in orthopedic surgery and had noted that titanium adheres strongly to bone and is not rejected by the body. This observation would eventually set Swedish professor of anatomy Per-Ingvar Brånemark on the road towards dental implants. Brånemark originally had no interest in dentistry; his focus was on the circulation of blood in bones. To get a glimpse at what was going on, he implanted a small tube equipped with a viewing lens into the leg bone of a rabbit. Familiar with the literature about titanium not triggering an immune reaction, he formulated his device out of this metal. When he eventually tried to remove the tube, he was unable to do so, the metal having thoroughly bonded with the bone.

At this point, Brånemark switched the focus of his research to the mechanism by which titanium fuses with bone and coined the term "osseointegration" for the phenomenon. He wondered about possible practical applications of this integration but first had to ensure that living human bone would not reject an implant. After all, people are not giant rabbits. Brånemark enlisted a number of volunteers from his research group who agreed to have a titanium screw inserted into a bone in their arm. There was no adverse reaction!

The stage was now set for a classic experiment. Given that the jawbone could be readily accessed, and that there were plenty of people struggling with dental problems, Brånemark now focused on titanium implants. Gösta Larsson, who had been born with a deformed jaw and had no lower teeth, volunteered to be a human guinea pig. In 1965, Larsson had four titanium implants surgically inserted to which artificial teeth were then attached. When he passed away forty years later, the implants were still in place, having transformed his life!

Dr. Brånemark went on to assemble a team of scientists to further investigate the interplay of bone and titanium. In spite of a number of publications attesting to the safety and longevity of titanium implants, the dental establishment remained unconvinced due to a long history of implant failures and the conventional wisdom that any foreign material would eventually be rejected by the body. Finally, after presenting conclusive evidence at a conference in 1982 in Toronto, the dental community began to embrace titanium implants from which by now millions have benefited.

Brånemark went on to expand his research beyond dentistry and made a mark in the use of titanium to attach prostheses to limbs and faces. Decades after becoming the first recipient of a "titanium dental fixture," Brånemark's preferred term, Gösta Larsson had a titanium hearing aid implanted!

Clearly, science often travels along unusual paths. Who could have predicted that an experiment with rabbits was a hop away from eliminating the embarrassment of a toothless smile?

IT STINKS!

"Gee, I don't know why people don't pick up after their dogs. It stinks here," moaned a friend as we were walking down a tree-lined street in Washington. The unmistakable scent of doggie doo was in the air, but heaps of the stuff would have been needed to produce such an intense odor. No such piles of poop were in sight, but there were plenty of what looked like squashed yellow cherries on the sidewalk. And they reeked! A mix of vomit, sweaty socks, and outhouse fragrance would be an apt description. An upward glance revealed that they were not cherries, but rather the seedpods of *Ginkgo biloba* trees, easily recognized by their duck foot–like leaves. Indeed, the ancient Chinese name for the tree, "yinxin," translates as "duck's foot."

The fleshy pods, or cones, which surround a nut are produced only by female trees and are not smelly until they are crushed. The nuts are edible and are used in Chinese and Japanese cooking, especially on special occasions such as weddings or New Year, but not in large amounts because they contain methylpyridoxine, a heat-stable toxin. When pods are trodden on, the terrifying smell of butyric acid is wafted into the air, and while this is repugnant to humans, it is believed that dinosaurs found the scent attractive and ate the pods, which led to the propagation of the trees as the seeds were expelled in the dinosaurs' excreta.

Yes, the trees were around in those days; botanists call the ginkgo a "living fossil" because it has existed virtually unchanged for some

200 million years. The tree is extremely hardy and demonstrated its survival strength in nineteenth-century London when the Industrial Revolution smothered the city in smog produced by the burning of coal. While people choked and trees withered, the ginkgos were unharmed. Even more impressive was the survival of a number of ginkgos in Hiroshima after the city was essentially destroyed by the atom bomb in 1945. Little wonder that many cities have taken to planting ginkgo trees on city streets, especially downtown, where other trees have a life expectancy of only a few decades.

The price to pay for the greenery is the putrid smell in the fall when the trees drop their cones that inevitably get crushed. Seoul, Korea, battles the scent by employing workers to hand-pick the cones before they ripen and fall to the ground. Some U.S. cities have taken to spraying the trees in the spring with chlorpropham, a herbicide that is commonly used to keep potatoes from sprouting, and in this case halts cone development. It is not a perfect solution because if there is rain after the application, the chemical gets washed away. Neither is planting only male trees since it turns out that ginkgos can spontaneously change sex as an evolutionary adaptation.

Butyric acid is one of the most obnoxious smells one can encounter. And you do not have to be stepping on ginkgo pods to be assaulted with it. Rancid butter or sweaty socks will do the job! Fats in butter react with oxygen in the air to yield butyric acid, the name of which derives from the ancient Greek for butter. The compound is also produced along with other delights such as dimethyl disulfide and dimethyl trisulfide that conjure up the fragrance of skunk when bacteria on the skin feed off the fats and proteins in sweat. Heating and air conditioning systems can also be plagued with what has been called "Dirty Sock Syndrome." This can be traced to mold and bacteria that grow when just the right conditions of moisture availability coupled with cycles of heating and cooling are met. Automobile AC systems will often release a smell when they are first turned on in the summer after only the ventilation and heating modes have been used during the winter.

Not all creatures are averse to the smell of butyric acid. Dogs are attracted to the smell, as are mosquitoes and bears. Not a good idea to wear smelly socks or T-shirts in the woods! An imitation foot odor scent, composed of ammonia, lactic acid, and fatty acids, including butyric, has been used to attract mosquitoes into traps and divert them from biting people, but it seems sweaty socks work better. The Bill and Melinda Gates Foundation funds an experiment in Kenya that involves collecting smelly sock odor with cotton pads and using these as bait in traps. Amazingly, the East African vampire spider that dines on blood-gorged mosquitoes is also attracted to butyric acid. That is an adaptation based on mosquitoes being attracted to the scent, so the spider knows that where there is butyric acid there will also be a meal.

While most people are repelled by smelly socks, some experience erotic arousal from the scent. This is a form of olfactophilia, or sexual arousal caused by body odor. Not surprisingly, many foot fetishists are turned on by smelly socks. Of course, being warded off by butyric acid is more common, not only in the case of people but in animals as well. Hoofed species, called "ungulates," such as horses, pigs, cattle, camels, deer, and sheep, stay clear of wolf, lion, and dog excrement because this signals the possible presence of predators.

In one study, the chemistry of dog feces was analyzed to investigate what component was responsible for the repulsive effect with a view towards using these to protect agricultural crops and trees from being feasted on by ungulates. One hundred and six compounds were isolated from dog feces, and it was determined that a mix of short-chain fatty acids, butyric being one, interfered with the appetite of sheep. When this mix was placed in the trough under the corn feed, a sheep delicacy, the animals either didn't eat or ate less.

Obviously, letting sheep loose on streets populated by ginkgo trees would not solve the smell problem. They would not dine on the dropped pods. Bears would be a better bet. On the other hand, they may prefer humans to ginkgo pods.

THOSE "FOREVER" CHEMICALS

Say "perfluoroalkyl substance." That's a mouthful, right? Luckily, we can get away with using the acronym PFAS, pronounced "peefas." But the real question is whether we are getting a literal mouthful of these chemicals when we eat or drink. Time to get an earful about PFAS.

For a start, there is general agreement that these chemicals are very useful but at the same time, very controversial. There was no controversy when 3M launched the first of these compounds in the 1930s as Scotchgard. Consumers were thrilled with the product's ability to stain-proof sofas, make carpets repel soil, and waterproof clothes. There was further joy in kitchens when polytetrafluoroethylene, trade named Teflon by DuPont, revolutionized cooking with nonstick surfaces. A further mark was made when PFAS were incorporated into firefighting. This happened after the 1967 *USS Forrestal* tragedy when a rocket was accidentally launched into armed airplanes on the carrier's deck and ignited a fire that killed 130 men. The firefighting equipment on board had proved inadequate, prompting researchers to develop a foam that would quickly smother fire. The key was the ability of PFAS to reduce the surface tension of water, allowing the foam to quickly spread over flames. Firefighting foams soon became standard equipment at airports and military installations around the world, saving lives.

Other applications followed. Perfluoroalkyl substances were ideal for cleaning chips in the electronics industry and grease-proofing food packaging such as pizza boxes and microwave popcorn bags. Today, some 1,400 of these chemicals are produced with over 200 applications, ranging from dental floss and ski wax to coatings on windmill blades and guitar strings.

The chemical feature that makes these uses possible, namely the strength of the bond between carbon and fluorine atoms, is also at the root of the controversy. It is the strength with which these atoms are held together that results in the environmental persistence of PFAS, leading to them being labeled as "forever chemicals." Indeed, they can be detected in our food, water, and, most concerningly, in our blood. Estimates are that roughly 95 percent of the population has measurable blood levels. Of course, the presence of a chemical cannot be equated with the presence of risk, but in this case, there is cause for some uneasiness.

The dark side of PFAS first came to light when cattle on a farm near a DuPont landfill became ill. The problem was traced to the animals drinking from a stream tainted by leachate from the landfill, prompting the farmer to seek legal counsel. Rob Bilott, a lawyer who took up the case, soon discovered an unusually high incidence of ulcerative colitis, kidney cancer, and birth defects in the area. He managed to turn up evidence that DuPont had failed to make public internal studies that had linked PFAS, particularly perfluorooctanoic acid (PFOA), used in Teflon manufacture, with toxic effects. This led to a class-action lawsuit resulting in a multimillion-dollar settlement and the creation of an independent panel of experts charged with examining the consequences of PFOA in the environment. A study of blood samples from 69,000 people living in the vicinity of the DuPont plant in Parkersburg, West Virginia, confirmed the presence of PFAS and also found a high incidence of kidney cancer, ulcerative colitis, thyroid cancer, pregnancy-induced hypertension, high blood cholesterol, and testicular cancer. The story is dramatically told in *Dark Waters*, a 2019 film that documents the legal battle between DuPont and Rob Bilott.

As evidence mounted against PFOA, including that it may be an endocrine disruptor, the chemical industry pledged to eliminate its use and managed to do so by 2015. After much research, the conclusion was that the problem was bioaccumulation in the body due to PFOA's lack of solubility. Once absorbed, the kidneys were unable to flush it out. This lack of solubility was traced to the compound's molecular structure, essentially to its basic skeleton of eight carbon atoms. Shorter chain PFAS to replace PFOA were developed and were claimed to be more soluble and less toxic. However, these are still environmentally persistent and whether they are actually safer is not clear.

And problems with PFAS just keep cropping up. Children with higher blood levels have been found to have fewer antibodies after diphtheria and tetanus vaccines. Even more alarming is that people with positive COVID tests are more likely to be hospitalized and go on to have serious consequences if they have detectable blood levels of perfluorobutyric acid (PFBA), the most widely used short-chain PFAS. Curiously, PFOA did not show such an association, underlining that PFAS cannot all be lumped together when exploring effects.

While there is no blazing inferno here, there is enough smoke to prompt efforts to reduce the public's exposure to these "forever" chemicals. In some applications, medical equipment for example, there are no ready substitutes, but elsewhere silicones can impart stain resistance, and polyester fabrics resist moisture without the need of any coating. Dental floss can be made without PFAS, as can cosmetics, carpets, nonstick cookware, and food packaging materials. Indeed, several states in the U.S. have already banned the use of these chemicals in materials that come into contact with food. And we don't need PFAS to lubricate bicycle chains or skis. Hydrocarbon lubricants do a good job.

Skiers, however, are loathe to give up fluorinated waxes because they reduce friction, and when race results are measured in fractions of a second, wax makes a difference. Nevertheless, the International Ski Federation is poised to ban fluorinated waxes for the 2023–2024 season, but that may stimulate cheating. Believe it or not, dogs are already being trained to detect the scent of fluorinated compounds and sniff out cheaters. But clandestine chemists are undoubtedly working on substances to mask the smell of fluorinated ski wax. By now I suspect you have had the earful I promised.

SILKWORM POO

"I don't want to gross you out, so instead of chicken guts I will use a piece of red silk." That's what I tell the audience as I begin my demonstration on "psychic surgery." Magicians, be they professionals or hobbyists, tend to have strong feelings about anyone who abuses their beloved art by fleecing the public. As "scientists of the stage," magicians use a variety of scientific and engineering principles to entertain audiences and have a strong disdain for charlatans who perform the same illusions but claim to do so by paranormal means. Fake psychics fall into this category, although I suppose the term "fake" is redundant.

Particularly offensive are the "psychic surgeons" who capitalize on cancer patients' desperation by pretending to remove tumors without making an incision. To a naïve observer this is very impressive as it gives the decided impression that diseased tissue is actually being removed. All that is required to carry out this amazing feat is a "special some-thing" and a little practice with sleight of hand. Many magicians, myself included, try to follow in the footsteps of James Randi, champion of critical thinking, by performing "psychic surgery" to demonstrate how easily people can be fooled into believing the unbelievable.

In my case, the role of a tumor is played by a small piece of red silk. I tell the audience that this "tumor" was produced by a worm, but in this case by a silkworm rather than a human worm. That description of a "psychic surgeon" usually gets a laugh. Although my mention of silkworms in this case is rather tangential, I have had a long fascination with the creatures. One of the first scientific toys I was ever given was a "Silk Factory" that came with silkworm eggs and instructions on how to hatch these into silkworms that would then spin cocoons to produce silk. I never managed to produce any silk, but the kit did produce an interest in silkworms and the moths they spawn. Much later, I was to learn of the fascinating chemistry that silk moths use to attract each other and wove this into a lecture I present on pheromones.

This preamble explains why I became so intrigued when I came across an item advertised for sale by a silk factory in Shanghai. "Silkworm excrement pillow" is the description of the unusual article and seems to be an accurate portrayal since the pillow is indeed filled with silk-worm poo. The poo, it is claimed, is "Chinese medicine" that "induces sweeter sleep for babies, relieves rheumatic pain in the aged, improves eyesight and promotes brainpower." While the idea of producing sweet dreams with a poo pillow may be novel, it turns out that the poo has a long history of use in Traditional Chinese Medicine. That is not too surprising because silk is woven into the fabric of Chinese culture.

According to legend, the production of silk dates back to around 2700 BC when Leizu, wife of Huangdi, the "Yellow Emperor," was having tea and a cocoon fell into her cup. In the hot water the cocoon unraveled, and the Empress was impressed by the long, strong threads that formed. She realized that the cocoon had fallen from a mulberry tree, and after some investigation learned that these trees were home to silkworms that spun cocoons from which moths eventually emerged. She encouraged the planting of mulberry trees to produce more silk fiber and even invented a loom that wove the threads into a soft fabric.

Production of the fabric was difficult, and silk became an important status symbol worn only by the nobility. It was so prized that sometimes silk was even used as currency, and anyone caught taking the secret of silk production out of China was supposedly put to death. Finally, around 550 AD, two monks managed to smuggle silkworm eggs out by hiding them inside bamboo walking sticks. Still, the Chinese managed to keep their grip on silk production and the trade route from Europe to China became known as the Silk Road.

Historically, Chinese Medicine has used all sorts of plant and animal products. The raising of silkworms generated lots of excrement and, it seems, a quest for potential uses. The poo soon wormed its way into medicine and claims began to circulate about its ability to "relieve rheumatism, dispel wind, improve sleep and strengthen the stomach." One would think it takes a strong stomach in the first place to swallow worm feces.

Could there be something to such worm therapy? Believe it or not, researchers have investigated the possibility resulting in publications with titles such as "Potential Pharmaceutical Uses of Isolated Compounds from Silkworm Excreta" and "Effects of Physiological Active Substance

Extracted from Silkworm Feces." The reference is to a number of flavonoids, sterols, and very long chain fatty alcohols that conceivably could have physiological activity. An interesting, albeit unsurprising, revelation. Mulberry leaves contain a host of compounds that can be then discharged in silkworm feces. Of course, such compounds are available from much more palatable sources, such as fruits and vegetables.

Now, about those pillows. While without further research one cannot rule out the possibility that some compounds are volatilized, the probability of benefiting nocturnally from laying one's head on a silkworm excrement pillow would appear to be very low. You can always give it a try by making your own. Silkworm poo is available on Amazon. If that doesn't appeal, you can try Sansha, a unique Japanese spirit brewed from mulberries and silkworm droppings.

Now for a final revelation. I'll let you in on a secret, but don't tell anyone. The "silk" I use in my demo is actually polyester. No silkworms were dropped into hot water in its production.

OXYGEN ON MARS

Landing humans on Mars is a colossal challenge. Returning them to Earth is an even bigger one. Launching a return vessel requires fuel and oxygen, and the amounts needed exceed what could be transported from Earth and would therefore have to be produced on Mars. Combustible methane could possibly be produced from the carbon dioxide that makes up roughly 95 percent of Mars's atmosphere, but supplying the oxygen needed poses an immense problem. The hope is that the Mars Oxygen In-Situ Resource Utilization Experiment that was delivered to the red planet aboard the Perseverance rover can serve as a prototype for the large-scale production of oxygen.

The curious name of the experiment was designed to yield the acronym "MOXIE," a word that has come to mean pep, daring, determination, and appropriately, perseverance. A fitting description for this incredibly complicated, spunky, and valiant effort. "Moxie" was introduced into our vocabulary back in 1876 as the name of a carbonated beverage concocted by homeopathic physician Augustin Thompson, who sought to create a medicinal brew that did not contain potentially harmful ingredients such as alcohol or cocaine, common at the time. His secret ingredient, said to come from a rare plant, turned out to be an extract of gentian root, named after Gentius, the first century BC king of Illyria, who reputedly used the herb as a tonic. Thompson marketed his product as a "most healthful drink that strengthens the nerves."

Indeed, it takes strong nerves to work on a project as complicated as NASA's MOXIE, that is based on a device roughly one cubic foot in size filled with innumerable valves, tubes, and electronics. The critical component is the solid oxide electrolyzer that splits carbon dioxide into carbon monoxide and oxygen. "Lysis" is the ancient Greek word for "breaking down," so that "electrolysis" means "breaking down with electricity." The term was coined in the nineteenth century by Michael Faraday, who conducted extensive investigations on the effects of passing an electric current through solutions. Anyone who has taken high school chemistry would be familiar with the classic electrolysis experiment of immersing a pair of electrodes in a container of water, connecting these to a battery, and watching bubbles of oxygen form at the positive electrode (anode) and hydrogen at the negative one (cathode).

Splitting carbon dioxide is more complicated than splitting water, but the principle is similar. The gas is introduced into an electrolytic cell that consists of a cathode and anode separated by an electrolyte, a substance capable of conducting electricity. In this case, the nature of the electrolyte is the key. It is a crystalline substance composed of zirconium oxide and yttrium oxide, similar to synthetic diamonds made of "cubic zirconia."

Carbon dioxide gas from the Martian atmosphere is first compressed, heated to a high temperature, then passed through a porous cathode, where it picks up electrons and breaks apart. The negatively charged oxide ions that form are then conducted to the anode through channels in the crystal lattice that are just the right size to allow these oxide ions and nothing else to pass. At the anode, the oxide ions give up their extra electrons to become atoms of oxygen that then join to form diatomic oxygen gas that can be collected and stored. The carbon monoxide that forms can also be collected and possibly used as fuel. Calculations show that a device about two hundred times the size of the current prototype would be capable of generating enough oxygen to sustain the astronauts on Mars and allow them to blast off.

What about the possibility of producing oxygen by the electrolysis of water? Not only can it be done, it is being done on the International Space Station. The space station is the ultimate example of recycling. All water is recycled, meaning that yesterday's sweat or urine become today's coffee. Of course, that happens after all wastewater goes through an elaborate purification system. Some of the water is then electrolyzed to produce oxygen and hydrogen, using electricity supplied by the giant solar panels. Enough oxygen is produced in this fashion for breathing. The hydrogen is combined with the carbon dioxide the astronauts exhale to produce water and methane. This water is added to the total water supply, and the methane is vented out into space. Since recycling is not perfect, some water eventually is lost so that periodically the station has to be resupplied with water ferried up by spacecraft.

Should the electrolysis system fail for some reason, an emergency oxygen supply is available from pressurized oxygen tanks, and when these run out, from chemical oxygen generators. These are based on a chemical reaction that is performed by many a high school student. Heating sodium chlorate in the presence of a catalyst such as manganese dioxide yields sodium chloride and oxygen. The space station is supplied with cannisters that contain the chlorate, the catalyst, and finely powdered iron. When needed, a percussion cap that contains a small amount of a shock-sensitive chemical sets off a mini explosion that ignites the iron powder that then burns and produces enough heat to decompose the sodium chlorate, releasing oxygen. This is the same chemistry used to supply emergency oxygen on airplanes.

Of course, such chemical generators could not meet the oxygen needs on Mars. However, water electrolysis could make a contribution if subterranean water exists in some form as is believed to be possible. For now, though, it seems that the In-Situ Resource Utilization Experiment that Perseverance successfully deployed may pave the way to landing people on Mars and returning them safely to Earth. There is no doubt that the scientists who designed the experiment were clever, creative, highly skilled, and determined. In other words, they had a lot of "moxie."

BULL TESTES

"You have to try them," the waiter replied when I asked about the "prairie oysters" on the menu at a Calgary restaurant. I was intrigued. I knew that ever since Pliny the Elder had recommended dining on hyena genitals soaked in honey for a "stimulating effect" some 2,000 years ago, the possibility of using testes in some form for rejuvenation had generated much interest. So, I thought, why not go for an unusual experience dining on bull testicles?

It had been apparent since antiquity that the loss of testes results in a loss of virility and fertility. Although as early as 1771 John Hunter had shown that a capon's manhood can be restored by re-implantation of its testicle, it wasn't until 1889 that an attempt was made to inject science into the role of the testes in human physiology.

It was then that physician Charles-Édouard Brown-Séquard at the age of seventy-two claimed that he felt rejuvenated after injecting himself with the testicular extracts of dogs and guinea pigs. Given that recent research has shown that his extracts could not have contained more than a trace of hormones, there is no doubt that Brown-Séquard had experienced the placebo effect. At the time, though, his "research" captivated not only members of the public who eagerly purchased Brown-Séquard Extracts produced by hucksters, but also stimulated other scientists to further explore the matter.

One of these was surgeon Serge Voronoff, who had studied medicine in France after emigrating from Russia in 1884 at the age of eighteen. He trained under Alexis Carrel, a surgeon who would go on to win the 1912 Nobel Prize in Medicine and Physiology for his method of suturing blood vessels. It was this technique that Voronoff would eventually put to a curious use after learning about Brown-Séquard's claim of regained youthful strength and sexual prowess.

Voronoff's experiments began upon returning to France after working as a physician in Egypt from 1896 to 1910. In Egyptian hospitals he had treated eunuchs and noted that these castrated men had flaccid muscles, lacked energy, and had memory problems. He now figured that Brown-Séquard's experience must have been due to chemicals produced by the testes and wondered if the need for repeated injection of extracts could be averted by implanting a testicle directly into the scrotum.

Voronoff started off by grafting the testicles of younger sheep and goats onto those of older animals and claimed restitution of youthful vigor. Realizing that human testicle donors would be hard to find, Voronoff managed to strike a deal with the French government for the testes of executed criminals. The positive publicity that Brown-Séquard's self-treatment had received ensured that there was no shortage of aging men who were willing to become recipients. Indeed, there were not enough men being hanged to meet the demand. The next best donors, Voronoff figured, were our cousins, the apes.

On June 12, 1920, Voronoff implanted slices of chimpanzee testes into the scrotum of a man and claimed a successful outcome. After a number of such operations, he presented his findings at the International Congress of Surgeons in London in 1923, to wide acclaim. Other doctors followed in his footsteps, creating a shortage in monkey glands that Voronoff remedied by starting his own monkey farm in Italy. He purchased a nearby castle where he performed his surgeries, charging an exorbitant fee for his "rejuvenation" procedure. However, the monkey gland fad soon faded as news began to spread about graft recipients continuing to age normally.

While Voronoff is the most famous of the physicians who carried out testicular implants, he was not the first to perform such surgeries. In 1915, George Frank Lydston in Chicago, also stimulated by Brown-Séquard's report, implanted testes from accident victims into older men. As early as 1904, he had experimented on animals, and before his pioneering surgery in 1915 on a patient, he had actually sewn a cadaver's testicle into his own scrotum with the assistance of a colleague. Lydston reported "a marked exhilaration and buoyancy of spirits." Unfortunately, he also believed in eugenics and thought that transplanted testicles could "cure" homosexuality.

At the same time that Lydston was experimenting with testicular implants, Leo Stanley, a physician at San Quentin prison, also carried out transplants from executed men into other inmates, even though he had no surgical experience. When he ran out of donors, he switched to using goat testicles, a procedure that John Romulus Brinkley, whose medical degree was from a diploma mill, would make famous. This "goat-gland doctor" performed more than 600 goat-to-man testicle implants before he was put out of business by numerous malpractice lawsuits.

The media reported widely on testicle transplants, leading some men to wonder if two testicles make a man masculine, would three not be better? A black market arose with accounts of men being chloroformed and falling victim to testicle larceny. Finally, the infatuation with these implants ground to a halt in 1935 when pure testosterone was isolated. This began the age of testosterone therapy, which is not without controversies of its own.

Now back to the "prairie oysters." The gustatory experience is not one I would ever care to repeat. As far as possible effects go, we know that testosterone is produced by the testes but they do not store the hormone. Furthermore, any orally ingested testosterone is inactivated as it passes through the liver. The bull would have gotten more use out of its testicles than I did.

BUT IT'S NATURAL!

Nature is amazing. Put a seed into the ground, and a plant emerges. Two cells unite, and a baby emerges. Two hydrogen nuclei in the sun fuse, and energy emerges. Take pollen from one flower, sprinkle it on a different type of flower, and a new variety of flower emerges. Is that flower natural? It would not have been produced if a human hand had not intervened. But isn't that hand also natural?

Why bring up this question? Because I have long been bothered by how the term "natural" is used. There are two issues here. First, the implication that natural substances are inherently safer than synthetic ones, and second, the rather imaginative stretching of the meaning of "natural." Certainly, "natural" does not equate to safe. Amanita muscaria mushrooms, tobacco plants, "poison dart" frogs, jellyfish, and snakes produce a large variety of natural toxins. We can add that bacteria, viruses, fungi, and parasites are also natural. And when a cell divides and accidentally produces a mutation in DNA that leads to cancer, well, that is also "natural." Indeed, much scientific research focuses on using synthetic preservatives and medicines to outwit nature in a decidedly unnatural fashion. The uncritical worship of "natural" is unwarranted.

How about the questionable use of the term? I recently came upon a lipstick advertised as "natural." Since I have never seen lipstick growing on a bush or tree, or secreted by some animal, I wondered about the justification of the term. Believe it or not, there actually is a "lipstick tree." Of course, it doesn't grow lipstick, but it does produce a colorant that can be used to formulate cosmetics. *Bixa orellana* is a shrub native to South America and Mexico that produces a stunning red fruit, the seeds of which yield a dye known as "annatto," with its major component being the carotenoid bixin.

Indeed, annatto may appear in lipsticks that then are advertised as "natural." Although adverse reactions to annatto are rare, they certainly have been noted in the literature. Allergies and exacerbation of irritable bowel syndrome can occur, and there is even a case report of a man suffering severe anaphylaxis with loss of consciousness within a few minutes of eating a sandwich with Gouda cheese. The use of annatto to give a yellow tint to Gouda is common.

It should also be mentioned that while the other components of such a "natural" lipstick, namely shea butter, castor seed oil, beeswax, coconut oil, and sorbic acid may also be found in nature, a great deal of processing is involved before they become part of a lipstick. Solvents are used for the extraction of the oils and waxes, and castor seeds have to be carefully processed to make sure there are no residues of the toxin ricin. While sorbic acid does occur naturally, it is actually produced industrially by reacting crotonaldehyde with ketene. Of course, that is irrelevant. The properties of sorbic acid do not depend on whether it was produced in the lab or in the unripe berries of the Himalayan rowan tree.

A greater stretch is with the claim of "naturally sourced." One of the most common ingredients in shampoos, toothpastes, and various cleaning agents is sodium lauryl sulfate (SLS), a compound that acts as a surfactant and detergent. Surfactants are substances that can insert themselves between water molecules and reduce the attraction between them. This allows water to spread more easily to "wet" a surface and also allows it to stretch around air bubbles to generate a foam. As for detergents, these are molecules with one end being attracted to oily substances and the other to water so that rinsing with a detergent solution can remove greasy soils. SLS is both an excellent surfactant and detergent, accounting for its widespread use.

Allegations that the compound is carcinogenic have no scientific basis, but the claim that it can act as a skin irritant has merit. The loud social media noise exaggerating the real and mythical risks has resulted in consumers shying away from products that contain SLS. This presents a challenge for industry because the ingredient is cheap and effective. One tactic has been to cloak it in a mantle of safety by suggesting that it is "naturally sourced." Yes, it is if you stretch the facts somewhat. Sodium lauryl sulfate is made by reacting lauryl alcohol with sulfuric acid. Lauryl alcohol is made by treating lauric acid with various reducing agents, and lauric acid in turn comes from the hydrolysis of trilaurin, a fat that is isolated from coconut oil. So, SLS is sort of "naturally sourced," but a lot of chemistry is involved in turning the coconut fat into the final product. Furthermore, "naturally sourced" has no relevance when it comes to safety or efficacy.

Another method to "hide" SLS was the subject of a *Wall Street Journal* article that accused actress Jessica Alba's Honest Company of being less than honest by declaring that their products were SLS-free. In response, the company retorted that their products are not formulated with SLS, but rather with "sodium coco-sulfate." Coconut fat actually contains several fatty acids that include stearic, oleic, myristic, palmitic, linoleic, capric, caprylic, and lauric acids, with the latter making up about 50 percent of the mix. When SLS is synthesized, the lauric acid is separated and sulfated, but it is also possible to sulfate the mixture and use that. Of course, this will then contain about 50 percent of SLS. The *Wall Street Journal* was correct, and the article triggered a "false ad" lawsuit that the company settled for one and a half million dollars.

It is about time to rid advertising of the "natural" babble and spread the message that the safety of a chemical is not determined by whether it is made by Mother Nature in a bush or by a chemist in a lab. It can only be revealed through proper scientific study. And remember that Mother Nature can be quite nasty. After all, viruses are natural. Vaccines are not.

GRAPHENE!

Sir Andre Geim, professor of physics at the University of Manchester, has the distinction of being the only person to have been awarded both a Nobel and an Ig Nobel Prize. The Nobel represents the pinnacle of scientific achievement, while the Ig Nobel, organized by the humor magazine *Annals of Improbable Research*, aims to "honour achievements that first make people laugh, and then make them think."

Geim received the 2000 Ig Nobel in physics for levitating a small frog with a powerful magnet and shared the 2010 Nobel Prize in Physics with Sir Konstantin Novoselov for isolating graphene, a remarkable material that is thinner, stronger, more flexible, and a better conductor of heat and electricity than any known substance.

To be sure, Geim and Novoselov did not "discover" graphene. Their Nobel citation reads "for groundbreaking experiments regarding the two-dimensional material graphene." Those groundbreaking experiments began with a piece of graphite, some Scotch tape, and an understanding that graphite is composed of planes of carbon atoms, each one bonded to three neighbors, 120 degrees apart, in a chicken-wire-like lattice. The planes can slide relative to each other, which explains why graphite is an excellent lubricant. An analogy would be a deck of cards, with each card representing a flat layer of carbon atoms. "Graphene" is the term for one such layer.

Now imagine using a piece of tape to remove one card from the top of the deck. This is just what Geim and Novoselov did with their piece of graphite and Scotch tape. Then they stuck the tape on a silicon substrate and removed it to leave a one-atom-thick sheet of graphene behind. Anyone can mimic this experiment using tape and the lead of a pencil, which is actually made of graphite. Make a mark on paper and apply a piece of tape on top. When the tape is pulled off, some of the graphite will have transferred to it. Take another piece of tape and stick it over the smudge on the first one and pull it off. Some graphite will have transferred to the second tape. Imagine repeating this process until you only have one layer of graphite left. You have just made graphene! But you are not the first to have made it.

In 1859, English chemist Benjamin Collins Brodie treated graphite with a strong acid and observed a suspension of tiny crystals. Not knowing anything about the atomic structure of graphite at the time, he thought he made a novel form of carbon. Actually, he had made graphene oxide, a graphene sheet with some oxygen atoms attached to it. Without realizing it, he had pioneered graphene research! The theoretical foundations for understanding the structure of graphene were laid in a 1947 paper by McGill University physics professor Philip Wallace, who was studying graphite, which at the time was of great interest because of its role in controlling the flow of neutrons in nuclear reactions.

Following up on Brodie's work, in 1962, German chemistry professor Hanns-Peter Boehm treated oxidized graphite with an alkaline solution and using an electron microscope identified what he suggested was "probably a single carbon honeycomb plane of the graphite lattice." This set the stage for Geim and Novoselov's isolation of pure graphene.

While the Scotch tape method can yield tiny amounts of graphene suitable for studying its properties, it is not amenable to the synthesis of large quantities. Excited by the material's potential uses, chemists and physicists all over the world were soon cranking out publications about graphene and various methods for producing significant amounts were discovered. Exposing methane gas to a super-heated sheet of copper results in graphene deposited on the copper, and using graphite as an electrode in an electrolysis reaction leads to tiny flakes of graphene being stripped off. Ultrasound and microwaves can also be used to "exfoliate" graphite to yield graphene.

Now that graphene can be made, the question is what to do with it. Ultra-long-life batteries, bendable computer screens, water desalination filters, improved solar cells, and superfast microcomputers are in the offing. So far, though, the only significant commercial applications have involved blending graphene with other materials to make stronger tennis racquets, hockey sticks, and tires, or so it is claimed.

However, with the outbreak of COVID-19, another use has emerged. Several types of masks that incorporate some form of graphene have appeared. In one variety, graphene is generated by exposing the non-woven polypropylene layer of a surgical mask to a laser beam. This results in an improved repelling of virus-bearing droplets and allows for quick sterilization of the mask by exposing it to sunlight. Another mask includes a layer of graphene oxide that has potent antiviral properties because it can physically sheer the lipid outer-coating of a virus, rendering it harmless.

This then brings up the question of whether such masks are harmless to the wearer. Can they release tiny particles of graphene that may be harmful if inhaled? Health Canada believes that is a possibility, at least with one type of mask that was distributed to schools and day-care centers. However, without details of Health Canada's findings it is not possible to evaluate the risks.

Certainly, inhalation of any particulate matter, especially smaller than 5 nanometers, can cause problems since on this "nano" scale, effects that might not be predicted based on bulk properties, such as triggering of an inflammatory response, may emerge. Consider, though, that there are many forms of graphene and of graphene oxide with different potential toxic effects. So, as far as the risk of the masks goes, let's recall Sherlock Holmes's famous dictum: "It is a capital mistake to theorize before one has data. Insensibly one begins to twist facts to suit theories, instead of theories to suit facts."

For now, it is appropriate to follow Health Canada's advice and avoid the masks in question, but let's not throw the baby out with the bath water. Indeed, if the movie *The Graduate* were remade today, the word whispered into young Benjamin's ear could well be "graphene" instead of "plastics." The material's potential is exciting, but light remains to be cast on the shadow of toxicity.

DUCT TAPE

U.S. President Franklin Roosevelt was intrigued by the letter he received from Vesta Stoudt, a woman who worked in a munitions plant where her job was to wrap the cartridges that were used to propel rifle grenades. The cartridges were packed in cardboard boxes that were then coated with wax to prevent any moisture from seeping in. A box was opened by pulling on a tab sticking out from a paper tape that had been used to seal the flaps under the wax. Mrs. Stoudt had noted that the tab was weak and would often tear off when tugged. This meant that soldiers would lose valuable time clawing at the wax to get their hands on the cartridges, giving the enemy time to approach. Since Mrs. Stoudt had two sons in the military she was concerned. And she had an idea!

In her letter to FDR, she offered a solution. Instead of the tape being made of paper, Mrs. Stoudt suggested it be made of strong fabric. Sounded simple enough, and the president forwarded the letter to the War Production Board that then put out a request for a cloth-based waterproof tape. That challenge was met by Permacel, a division of Johnson & Johnson, a company that already had some expertise in developing medical tapes. To seal the ammunition boxes, scientists came up with a layer of tightly woven fabric sandwiched between a rubber-based adhesive and a coating of polyethylene. The fabric used was "cotton duck," the name having nothing to do with waterfowl. It was derived from "doek," the Dutch word for a type of canvas that was used to make sailors' garb. Polyethylene, a plastic introduced by Imperial Chemical Industries in the 1930s, was the key to making the tape waterproof. It wasn't the only role this polymer played during the war. Polyethylene was a critical insulating material used in the construction of radar equipment, its light weight allowing airplanes to be equipped with radar.

Soldiers found that the new tape had uses beyond sealing ammunition boxes. It came in handy when repairs were needed for all sorts of equipment and even was pressed into use for wounds when nothing else was available. Some historians have floated the idea that soldiers, either because they were aware of the "cotton duck" connection or because they thought the tape shed water like a duck, began to use the term "duck tape." It is difficult to find evidence for this, but it is clear that in 1975 the Manco company obtained a trademark for "Duck

Tape" together with a yellow cartoon duck logo and explained this was a "play on the fact that people often refer to duct tape as 'duck tape.'" The term "duct tape" had come into common use after the war when manufacturers of heating and air conditioning ducts discovered that the tape was useful for connecting components. The color was changed from the original military olive green to silver by using aluminum powder so as to blend in with the tin ducts.

Whether it was called "duck" or "duct," numerous clever uses were soon found for the tape. During the Vietnam War, holes in helicopter blades caused by enemy fire were temporarily repaired with it, and in 1970 the tape saved the lives of the three Apollo astronauts who had to use the lunar module as a "lifeboat" after the explosion of an oxygen tank. Duct tape was critical in modifying the command module's square carbon dioxide filters to fit the lunar module's round receptacles. In 1972, *Apollo 17* astronauts Harrison Schmitt and Eugene Cernan managed to repair a fender on their lunar rover with duct tape.

Then in 1998 came a stunning bit of research. Max Sherman and Iain Walker of the Lawrence Berkeley lab in California found that duct tape "failed reliably and catastrophically" in preventing energy loss when used in the heating and cooling ducts of houses. As one might expect, the revelation that duct tape should not be used on ducts caused quite a media frenzy. This had the beneficial spin-off of consumers asking that their ducts be sealed with superior materials, known in the trade as "mastics." These can be formulated with various acrylics, silicones, or synthetic rubbers and lead to considerable energy savings. Indeed, building codes have been changed to prevent the inappropriate use of duct tape.

Of course, duct tape remains popular for other applications. It has been used to hold broken tail lights in place, secure wigs, and even remove warts. A much-publicized research paper in 2002 reported that duct tape applied to a wart for two months was more effective than the usual treatment with liquid nitrogen, although subsequent studies failed to corroborate this finding. A judge in Ohio once ordered an abusive defendant's mouth to be duct-taped shut during a trial, and criminals, at least in movies, have used duct tape to bind their victims to chairs. A clever quote attributed to one G. Weilacher has been making the rounds. "One only needs two tools in life; WD-40 to make things go, and duct tape to make them stop." An ingenious woman once stopped her husband from leaving the toilet seat up by taping it down with duct tape. Hope he had good aim.

PORCELAIN AND ALCHEMY

Rumpelstiltskin was the tricky little imp with the remarkable ability of weaving straw into gold as described in a classic Grimm Brothers fairy tale. Could this story, first published in 1812, have been stimulated by a real-life crafty alchemist who claimed to have accomplished the elusive quest of turning ordinary materials into gold?

Sometime in the early 1700s, young Johann Friedrich Böttger was apprenticed to a Berlin apothecary. At the time the search for medicines was intermingled with the pursuit of the "philosopher's stone," the legendary mineral that when combined with base metals would produce gold. Even famous scientists such as Isaac Newton and Robert Boyle dabbled in alchemy but came up empty. Böttger, however, was convinced that the secret lay within his grasp if he could just raise funds for his experiments. Money would surely flow in, he thought, if he could demonstrate that he was on the right track. As he was quite adept at conjuring, he managed to fool the gullible about already having discovered the secret of transmutation by using sleight of hand to make a gold nugget appear in his crucible.

Stories of Böttger's exploits reached the ears of "Augustus the Strong," ruler of Saxony and King of Poland. Hungry for gold, Augustus had Böttger arrested and locked away with an array of books and alchemical apparatus, tasked with producing the precious metal. After years of failure, and with the threat of execution hanging over his head, Böttger hatched a scheme to save his neck. He admitted that his search for the philosopher's stone had been futile but professed that in the process he had discovered another elusive secret, that of making "white gold," as porcelain was known in those days. This intrigued Augustus, who had become infatuated with the substance, the discovery of which traces back to the eighth-century Tang Dynasty in China. (Hence the term "china" to describe porcelain.) Since all European attempts to reproduce the Chinese process had been unsuccessful, the king decided to give Böttger a shot at showing what he could do. To ensure that he was not being duped, Augustus assigned Ehrenfried Walther von Tschirnhaus, the leading Saxon scientist at the time, to supervise the project.

Tschirnhaus had worked with Newton and Boyle and had himself been captivated by the prospect of making porcelain. Unlike European pottery, porcelain was white, translucent, and nonporous. Introduced into Europe by Marco Polo, the glass-like substance was named "porcelain" from the ancient Italian "porcellana" for a type of sea snail that has a smooth and shiny shell. It was assumed that, like all pottery, the production of porcelain involved subjecting clay to heat. But the type of clay, the temperature needed, and any other components that may be required were a mystery to European potters.

In his alchemical experiments Böttger had used special durable stoneware crucibles and had noted that when he heated certain minerals in them, the crucibles themselves developed a whitish sheen. Indeed, it was likely this observation that gave him the idea of attempting to make porcelain. Tschirnhaus had tried to discover the secret of porcelain by melting down a Chinese sample to see what its components may be. To produce the high temperature needed, he designed a series of lenses that focused sunlight on the sample, having learned how to grind lenses from none other than Christian Huygens, the Dutch polymath who revolutionized the making of telescopes.

Böttger and Tschirnhaus now began heating different kinds of clay at temperatures never previously attained by European kilns and adding various substances called "fluxes" to promote melting. Eventually they hit upon a type of clay called kaolinite, named after Kao-ling, the region of China where it was originally found, that when mixed with alabaster (calcium sulfate) acting as flux, produced a hard porcelain-like material. Due to iron impurities, it was reddish and would come to be known as "Böttger stoneware." Success eventually came when a slurry of kaolinite, quartz (silicon dioxide) and a flux of feldspar (a potassium aluminum silicate mineral) was poured into a mold and fired first at a lower temperature then at a very high temperature. Augustus was thrilled and set up a factory in the town of Meissen to produce the first European porcelain figurines and dinnerware that were comparable to Chinese imports.

The king amassed a collection of some 30,000 pieces of "Meissen" as well as Chinese porcelain, about 8,000 of which survive and are on exhibit at the Zwinger Museum in Dresden. Some Meissen pieces are very valuable; a teapot that once belonged to Princess Sophia of Hanover, mother of King George I of England, is valued at more than $320,000! Regrettably, Friedrich Böttger never saw profits from his work. Although Augustus gave him his freedom on condition that he would not reveal the secret of porcelain, he died in debt at the age of thirty-seven, likely poisoned by all the toxins he had inhaled in his quest for riches. Next time you serve dinner on a porcelain plate, or marvel at your Lladró, Royal Doulton, Herend, Royal Dux, or, if you're lucky, Meissen figurines, think of Böttger, the conjuring scoundrel whose alchemical experiments resulted in solving the mystery of "white gold."

LEAD — IT REALLY IS TOXIC

French physician Louis Tanquerel des Planches made an interesting observation in the early nineteenth century. Naval officers on some ships complained of muscle aches and abdominal cramps while ordinary sailors were spared this affliction. It turned out that officers' cabins were routinely painted but the sailors' quarters were not. White paint at the time was usually formulated with linseed oil and lead carbonate, and the officers were showing the symptoms of lead poisoning.

Dr. des Planches was able to make the connection because he had noted similar symptoms in a number of his patients at Hôpital de la Charité in Paris. A common feature among these patients was that they had some sort of exposure to lead, either through paints, cosmetics, food, beverages, or work. Lead carbonate was used by noblewomen to give their face a white appearance, candies were sometimes colored with lead chromate, and beverages were consumed from pewter mugs. Water and sewage pipes made of lead were commonly manufactured, exposing workers to the metal.

Des Planches put it all together in 1839 in a classic treatise on the "Saturnine Disease," as he called lead poisoning. That term derives from the ancient Roman god, Saturn, who was a demonic, irritable god. Irritation of the gut, a condition that afflicted many Romans, came to be referred to as "saturnine gout." The cause was likely lead poisoning since the Romans used lead pipes and lead dinnerware. Indeed, Vitruvius, Julius Caesar's engineer, noted that "water is more wholesome from earthenware pipes than from lead pipes" and warned against the use of lead. Hippocrates had earlier observed lead poisoning among miners, and Dioscorides, a Greek physician employed by the Roman army, wrote that lead makes "the mind give way," an observation that meshes with the current understanding that lead toxicity affects the brain.

Lead may even have been the reason that Van Gogh's mind "gave way." There is no question that the artist suffered from psychiatric problems and was fond of using yellow lead chromate, as in his famous sunflowers. He often licked his brushes to straighten the hairs! Other cases of mental problems attributed to lead toxicity speckle history. In the eighteenth century, "Devonshire colic" in England was likely caused by drinking cider made in presses lined with lead. Beethoven may have suffered from lead poisoning by drinking cheap wine, illegally sweetened with lead acetate. Thus, there is no doubt that by the time tetraethyl lead (TEL) began to be added to gasoline in the 1920s, the toxicity of lead was well known.

Early gasoline engines "knocked" due to uneven burning of the fuel in the cylinder. Since this reduced the power of the engine, a search was on for a remedy. General Motors engineer Thomas Midgley tried camphor, ethyl acetate, aluminum chloride, iodine, ethanol, and even melted butter as "anti-knock" gasoline additives. Ethanol worked well, but it couldn't be patented so GM wasn't interested, and petroleum companies were concerned that adding ethanol to gasoline would lead to a reduced need for petroleum.

Midgley persisted and finally came up with tetraethyl lead, a substance discovered back in 1854. It prevented knocking admirably. This interested GM, but there was the thorny issue of toxicity. Facing a group of skeptical reporters at a press conference, Midgley tried to allay concerns by washing his hands in a solution of tetraethyl lead. A curious demonstration, given that he himself had spent several months recuperating from symptoms that smacked of lead poisoning. The very week that Midgley carried out his infamous hand washing, five workers died at the Standard Oil plant where tetraethyl lead was being produced. There had also been reports of workers hallucinating, and one of the buildings where researchers had been experimenting with tetraethyl lead came to be known as the "looney gas building." At a production plant in New Jersey, workers kept hallucinating about insects and the facility became known as "the house of butterflies."

The leaded gasoline controversy led to the government organizing a conference in which the vice-president of the Ethyl Corporation that had been formed to produce the additive referred to tetraethyl lead as a "gift from God." Dr. Alice Hamilton, the country's foremost authority on lead, argued that the devil was a more likely proponent, claiming that "where there is lead, sooner or later cases of lead poisoning develop."

Regulators ended up siding with industry, claiming that a low level of lead can be tolerated, and by the mid-1920s leaded gasoline had become the primary fuel for automobiles. Nevertheless, research into possible toxic effects continued, and by the 1970s studies had shown that average blood lead levels had risen sharply, especially in children, and that people who lived close to highways had higher levels. Some scientists even linked rising crime rates to leaded gasoline.

A phaseout finally began in the 1970s, although it wasn't due to people being poisoned by lead. It was catalytic converters that were being poisoned! These had been introduced to meet stricter tailpipe emission controls, but the catalyst could not stand up to lead. This, together with mounting concerns about toxicity, resulted in tetraethyl lead being banned by the 1990s, replaced with ethanol and other additives that had been developed

When TEL was first removed, some people protested that the reformulated gasoline was less powerful and they wanted the old "leaded" variety back. This even led to the word "leaded" taking on the meaning of "powerful" and being used as slang for dark, strong coffee. "Don't give me any of that anemic, unleaded decaf," some would say, "give me the robust 'leaded' kind." This even spawned a company called Leaded Coffee.

When a 2021 study documented an association between dark coffee consumption and a reduced risk of heart failure, many media reports began with "drinking one or more cups of plain, leaded coffee a day is associated with a long-term reduced risk of heart failure." Don't worry, it is just an expression, there isn't any lead added to coffee.

OH, THAT OLD BOOK SMELL!

"I love the smell of book ink in the morning," exclaimed Italian philosopher and novelist Umberto Eco, who certainly knew a fair bit about books. Not only did he write more than fifty, but in one of his most famous works, *The Name of the Rose*, a book plays a central role. It is the murder weapon!

In this medieval mystery, a monk in a monastery is disturbed when he learns that some of his associates are reading an ancient text by Aristotle that he believes undermines faith in God. This, he thinks, is a crime punishable by death. Knowing the habit of moistening the fingers to turn a page, the murderous monk taints the pages of the book with arsenic and successfully dispenses with those who in his eyes are "heretics." A Franciscan friar, William of Baskerville (Eco is obviously a fan of Sherlock Holmes), exposes the killer and his revelation results in the villain committing suicide by consuming the poisoned pages of the book.

Could such a case of "murder by book" really happen? That isn't as far-fetched as it may sound, since arsenic compounds can be toxic at very low levels. While *The Name of the Rose* is totally fictional, at the University of Southern Denmark, a research librarian, with help from a chemistry professor, really did discover significant amounts of arsenic on the covers of three books dating from the sixteenth and seventeenth centuries. Some of the lettering had been obscured by a mysterious green color, and in an attempt to read the script, the books were subjected to a type of x-ray analysis. Surprisingly, this revealed that the green pigment was copper acetoarsenite, also known as "Paris Green."

Is it possible that arsenic could have been used to taint the books with a poison in a fashion similar to what Eco's mad monk had done? Although this cannot be totally ruled out, it is far more likely that the arsenic compound was affixed at some later date to protect the books against insects and vermin. "Paris Green," which derives its name from efforts to dispatch rats infesting the Paris sewers, was widely used as a pesticide well into the twentieth century. It could have been added to book covers to deter insects and rodents from dining on them. Plants are certainly on the menu for pests, and since paper is made from either papyrus, flax, bamboo, cotton, or trees, then as far as bugs are concerned, any paper scent may be akin to ringing the dinner bell.

Anyone who has meandered through a used bookstore, or has opened a freshly printed book, will attest to books having a characteristic smell. Until the nineteenth century, paper was made mostly from flax or cotton fibers, which are composed essentially of cellulose. With time, on exposure to air, particularly if the air contains traces of acids, cellulose can undergo a number of chemical reactions that lead to the release of a variety of fatty acids, alcohols, and aldehydes, all with distinct odors. These smells, along with those released by molds, such as the musty fragrance of trichloroanisole, contribute to that "old book smell." Furfural, a breakdown product of cellulose, adds an almond-like fragrance, and interestingly can be used to determine the age of a book. It is found in higher concentrations in books published after the mid-1800s, when pulp from trees replaced cotton or linen paper.

Pulp, like cotton or linen, is basically cellulose, but it also contains lignin, a complex cross-linked phenolic polymer that decomposes to yield a number of compounds with vanillin, the characteristic smell of vanilla being one of them. Since lignin also releases acids, it will increase the amount of furfural formed from cellulose, which is why the amount of furfural detected can be used to determine if a book was printed before or after the introduction of wood pulp. Paper made from wood pulp had its furfural content further enhanced when the combo of rosin, a resin obtained from pine trees, and aluminum sulfate was added to reduce absorbency and minimize the bleeding of inks. This made the paper more acidic, leading to enhanced degradation of cellulose and increased formation of furfural.

Modern paper mostly uses pulp from which lignin has been removed by a chemical process, except for cheaper varieties like newsprint, which will consequently degrade and yellow more quickly. Lignin-free paper has fewer degradation products, but chemical pulping uses the likes of sodium bisulfite that can release smelly sulfur compounds. Then there are the adhesives used in binding, vinyl acetate-ethylene copolymer being one, alkyl ketene dimer to prevent water absorption, and bleaching agents such as hydrogen peroxide or chlorine dioxide.

Add to these the solvents for ink, and you have the cacophony of compounds that make up a "new book" smell. This is a scent that some people miss when they read electronic books! Inventors have risen to the occasion, producing a host of candles and sprays that mimic "book scent." Kindle readers can now revel in sniffing "odeur de livre" as they try to follow the complex plot in *The Name of the Rose*. And of course, they don't have to worry about any possible exposure to "page-turning" poisons.

ROOTS OF FRENCH WINE

It seems the toads and chickens didn't have much of an appetite for aphids, the tiny insects that suck the nutrient-rich sap out of plants and are the bane of horticulturalists. When *Phylloxera vastatrix*, a species of aphid, infested French vineyards in the latter part of the nineteenth century, some desperate vintners positioned toads under the vines, hoping the creatures would dine on the little bugs that threatened to destroy the wine industry. Others allowed chickens to roam through the vineyard, thinking they would peck away at the insects. Neither approach worked. The insects also defied the pesticides available at the time. Finally, French winemakers Leo Laliman and Gaston Bazille found a solution to the phylloxera infestation and put French wines back on French tables to the relief of the population.

Laliman and Bazille were familiar with accounts of the failure of French colonists in America to grow grape vines they had brought from Europe. The colonists did not know why the vines failed to thrive but noted that there was no problem growing native American grapes and switched to these albeit unhappily because they thought French grapes produced better wines. A possible explanation for the failure of the French vines to grow in America arose in 1870 when American entomologist Charles Valentine Riley confirmed that phylloxera destroyed the roots of vines by injecting a venom that allowed them to feast on the sap.

Could it be that American rootstock had evolved resistance to the insect's venom, but French vines had not since the phylloxera aphids were not native to Europe? Laliman and Bazille now had an idea. Import American rootstock and graft shoots of French vines onto them. That worked! The shoots grew and developed into vines that produced the desired grapes. While American roots saved the French wine industry, there was some poetic justice here since it was the importing of American plants that accidentally introduced phylloxera to Europe in the first place.

Phylloxera was not the only problem introduced by American botanical specimens. Various fungi capable of causing plant diseases also made their way across the ocean. One of these was downy mildew, a fungus that stunts the growth of plants. The Bordeaux region of France was particularly affected, and vintners sought help from University of Bordeaux botany professor Pierre-Marie-Alexis Millardet. As he walked through the vineyards, the professor made an interesting discovery. Vines that bordered the property close to roads showed no mildew, while other vines were affected. Upon questioning the growers, he learned that they had a problem with passers-by helping themselves to the grapes. The losses were significant enough to take action, which they did by spraying the vines with a mixture of copper sulfate and lime (calcium hydroxide). This results in a precipitate of copper hydroxide that imparts a bitter taste to the grapes and also has an unappetizing green color. Millardet wondered if this was also the reason for the lack of mildew on these plants. Trials with this combination of chemicals soon revealed that they prevented fungal spores from germinating, and that spraying with this "Bordeaux mixture" pre-empted the fungal disease. French palates were mightily grateful for Millardet's investigations of vintners' efforts to deter grape thieves.

Not all fungi are detested by winemakers. *Botrytis cinerea*, better known as "noble rot" is welcomed by producers of dessert wines such as Sauternes from Bordeaux, some Rieslings from Germany, and perhaps most famously, Tokaji Aszú from Hungary. Grapes infected with this fungus dehydrate and shrivel, meaning that about a kilo of grapes is needed to yield a few milliliters of juice. Since the fungus does not affect the sugar content, the resulting wine is very sweet. But it is not only the sweetness that is prized, it is the flavor that the fungus contributes, some of which is attributed to compounds such as phenylacetaldehyde produced by the fungus.

Botrytis requires very special weather and soil conditions to thrive, and furthermore, since not every grape is affected, the shriveled ones have to be hand-picked. Because of such factors, botrytized wines are expensive, with one of these, a 2008 vintage of Essencia from the Tokaj region of Hungary that is fermented for some eight years, being the most expensive wine in the world. It retails for an astounding $40,000 a bottle! This gem is to be sampled by oenophiles only a spoonful at a time, and according to some, on one's knees, so that appropriate homage is rendered to its incredible quality, which is said to be maintained for 200 years. Maybe around the year 2200, some clever columnist will be able to comment on this being put to a test.

Although I have not been privy to this holy experience, I have visited Tokaj and have sampled the less exalted but still world-famous Tokaji Aszú, also made with nobly rotten grapes, said to have been called the wine of kings and the king of wines by none other than Louis XIV. To be honest, that is not why I was drawn to sample a bottle in Tokay. It was because Sherlock Holmes once offered a glass to Watson, saying, "A remarkable wine, Watson. I'm assured that it is from Franz Joseph's special cellar at the Schonbrunn Palace." If Holmes recommended it, that was good enough for me! As if that weren't enough motivation, I also knew that in the book *The Phantom of the Opera*, the Phantom offers Christine a glass of Tokaji on her visit to his lair below the Paris Opera.

And how did the Tokaji Aszú taste? I can say it was a delightful complement to the veal paprikash and langos we were served. Blessed be to the fungus!

LET'S PLAY CHESS

I learned about the Queen's Gambit long before Netflix's miniseries by that name became wildly popular. In 1955, in Hungary, then solidly behind the Iron Curtain, we had chess lessons in class. The Soviets believed that proficiency in the world's most popular board game could be used to demonstrate the intellectual superiority of communism. Of course, in grade three, I knew nothing about the Cold War and just enjoyed learning to play the game. After coming to Canada, my interest increased because reading chess publications did not require a knowledge of English.

It didn't take long to learn the language, and I was soon reading about a young phenom in the world of chess who had won the U.S. championship in 1958 at the remarkably young age of fifteen. Bobby Fischer would go on to repeat this performance seven more times! I was tickled pink when I heard that he would be coming to Montreal in 1964 to play in a simultaneous exhibition against fifty-six experts at Sir George Williams University, now Concordia. I was there to watch him beat forty-eight of the opponents, draw with seven, and lose to only one. The next day at the Jewish Public Library he defeated all ten opponents in another brilliant display of simultaneous chess. Fischer seemed affable, witty, and had a dry sense of humor. I liked him.

I didn't hear much more about Bobby until an eventful September weekend in 1972, although I would later learn that he had appeared on U.S. talk shows criticizing the Soviets for rigging tournaments and had bluntly stated that he was ready to challenge for the World Chess Championship. And challenge, he did! In preliminaries Bobby blew away two major contenders with unprecedented scores of 6–0, earning the right to play the then current champion, Boris Spassky, in Reykjavík, Iceland.

The prospect of this match at the height of the Cold War was portrayed as an intellectual showdown between the "communists" and the "free world." For the first time ever, a chess match would be televised live! Chess fever in the U.S. skyrocketed. Unfortunately, there was controversy from the beginning, with Fischer refusing to appear in Reykjavík for the opening match unless the prize money was increased. When he finally did arrive, after a patriotic appeal from Henry Kissinger, he complained that the spectators were too close and the television cameras made too much noise. He would not play unless the situation was remedied. The organizers finally gave in to the demands and the games were played in a room with no spectators and equipped only with closed-circuit cameras.

In what came to be called "The Match of the Century," Fischer defeated Spassky, ending twenty-four years of Soviet domination, becoming a hero to Americans and to me as well. I eagerly read reports of the match and was elated when Spassky conceded the final game on September 1. My happiness would not be long-lived because the next day I was at the Montreal Forum to watch the Soviet hockey "amateurs" demolish the Canadian "professionals" in the first game of another "Match of the Century."

Having your heroes knocked off a pedestal is very disturbing. While the Canadian pros did come back to defeat the Soviets in hockey, Bobby Fischer would soon begin to teeter on his pedestal. In 1975, he was due to defend his title against young Soviet sensation Anatoly Karpov, but it was not to be. This time the International Chess Federation would not meet Fischer's ever-increasing outrageous demands, and he was forced to forfeit the title. He then disappeared from public view until 1992, when he emerged from seclusion for a rematch with Boris Spassky in Yugoslavia that he again won. By this time, though, the press had been reporting on Fischer's eccentricities and paranoia. He believed that television sets were emitting dangerous radiation, that the Soviets were trying to poison him and were monitoring his activities by reflecting radio signals from his dental fillings. He had the fillings removed. Bobby had joined the Worldwide Church of God, whose leader Herbert Armstrong prophesized that the world was soon coming to an end. When this was unfulfilled, Bobby left the church and attacked it vigorously for being "Satanic."

It was after the 1992 match that Fischer not only figuratively toppled from the pedestal, but crashed and burned. The U.S. had banned the conducting of any sort of business in Yugoslavia because of the war it was waging against Bosnia and warned Fischer that if he played Spassky he would be arrested if he returned to his homeland. As recorded by television cameras, he spit on the letter, beginning a crusade of hatred for America and particularly of Jews, whom he said were running and ruining the country. Bobby Fischer became a raging anti-Semite despite his mother and biological father being Jewish. He applauded the 9/11 attack on the Twin Towers and called for synagogues to be closed down and the "execution of hundreds of thousands of Jewish ringleaders."

Banned from the U.S., he lived for a while in the Philippines, Hungary, and Japan. In 2004, he was detained in Japan for trying to leave the country with a U.S. passport that had been revoked. The Icelandic chess community came to his rescue by petitioning the government of Iceland to grant him asylum and citizenship, which it did. Fischer lived in Iceland until his death from kidney disease in 2008, withdrawing from chess, but not from his lunatic rants against Jews and America. His mental condition has been discussed in the medical literature, but since Fischer never had a formal evaluation by a psychiatrist, only theories have been offered, with paranoid personality disorder being the most prevalent.

Bobby Fischer died without a will, leaving his American nephews; Japanese wife, Miyoko Watai; and Marylin Young, a Filipino woman who claimed that her daughter Jinky had been fathered by Fischer to battle for his estate, valued at around $2 million. A petition by Young to have Fischer's body exhumed was accepted by Iceland, but analysis of the DNA samples clearly ruled out Fischer being Jinky's father. A court concluded that Miyoko Watai, a pharmacist and the president of the Japan Chess Association, had indeed married Mr. Fischer in 2004 in Japan and was therefore entitled to inherit his estate.

Bobby Fischer left an enigmatic legacy. He was undoubtedly one of the most brilliant chess players of all time, but his accomplishments will forever be clouded by his descent into the pit of racist paranoia.

Now getting back to those chess lessons the Soviets made part of our elementary school curriculum. Maybe they were really on to something. A number of studies since the 1970s have demonstrated that chess enhances both critical thinking and creativity. And given the number of people today who believe that Earth is flat, that vaccines contain microchips, that the moon landing was faked, that dowsing rods can detect water, or that metal spoons can be bent by mind power, we are in sore need of critical thinking.

Critical thinking involves coming to a conclusion about a question or problem that is based on the proper interpretation of observations, differentiation of facts from fallacy, awareness of the laws of nature, and familiarity with previous explorations of the subject. How can chess possibly help with this? Just think about it. Chess is essentially a game of identifying a problem and trying to solve it within a set of rules. It involves learning from experts, learning from one's mistakes, and analyzing the possible outcomes of a move. That's critical thinking. But is there any scientific evidence that chess can foster this? There is.

Evaluating critical thinking is not easy, but there are useful tests such as the one first developed by psychologists Goodwin Watson and Edward Glaser in 1925 that has had numerous revisions. Using a multiple choice format, Watson–Glaser tests are designed to measure the ability to draw inferences from data, to recognize the validity of an assumption, and to determine whether a given conclusion can logically be drawn from an argument. In a classic study in the early 1980s at a Pennsylvania high school, students who were enrolled in a weekly chess class improved their test scores significantly when compared with a group of control students.

A 1992 Canadian study in a New Brunswick elementary school compared students who were taught math in the traditional fashion with ones whose math course was enriched with chess. These students performed significantly better on problem-solving tests! Chess benefits are not limited to students. A 2021 study of seniors in Spain showed that two sixty-minute sessions of chess training per week resulted in improved cognition when compared with a control group, and participants even claimed an improvement in quality of life. And there is no downside to playing chess!

Now it's time to set up the chess board. But not exactly the way I did it back in elementary school. Times have changed, and now I play online with my grandson who is halfway around the globe. Maybe I'll try the Queen's Gambit.

INDEX

Kemmerich, Edward, 65
keratin, 139
keratin treatment, for hair, 138–39,
 140–41
ketamine, 110
Kevlar, 37–38
Keys, Ancel, 95
Knowles, Stanley, 135
Knowles, William, 109
"kupfernickel," 58–59

lactobacilli, 33
Laliman, Leo, 204, 205
Lampe Berger, 12–13, 15
Larsson, Gösta, 173, 174
"laughing gas" (nitrous oxide), 11, 56,
 146
Lavoisier, Antoine, 1–2, 85, 115
L-Dopa, 109, 135
lead
 in crystals, 132
 in gasoline, 200–201
 molten lead "trick," 78, 79
 for pellets, 67–68
 toxicity and warnings, 72, 74,
  199–201
lead carbonate, 199
lead chromate, 164, 199
"lead glass," 131
lead poisoning, 199, 200
Lebedev, Sergei, 23
Leidenfrost, Johann Gottlob, 78–79
Leidenfrost effect, 78–80
Leizu (empress), 181
Leucippus, 53
Liebig, Justus von, and Liebig
 Company, 27, 64–66
lie detectors, 79
Lifson, Nathan, 2
light, 75–77, 114
lighters, 14
lignin, 203

Lime, Harry (movie character), 50, 51
lipstick, as "natural," 188–89
"lipstick tree" (*Bixa orellana*), 188
List, Benjamin, 120, 121, 122
Livry, Emma, 8
Long, Crawford, 57
longevity, 105, 168
"Love Chair," 80
low-temperature cooking, 160,
 161–62
L-tryptophan, 101
Luddites, 7
Luxardo, Giorgio, 159
Luxardo distillery, 157, 159
Lycra, 129
Lydston, George Frank, 187

MacDougall, Duncan, 118
MacMillan, David, 120, 121, 122
magic tricks and magicians, 169, 180
Maillard, Louis Camille, 162
Maillard reaction, 162
malaria, 123–24
Marasca cherries, 157, 158
maraschino cherries, 157–60
Marcel, Jesse, 170
margaric acid, 19
margarine, 18–20, 97
Mars, 182, 183–85
Marsden, Ernest, 53
Martin, A.J.P., 5
McGuff, Paul, 76
McMath, Jahi, 117
meat
 cooking techniques, 161–63
 juice and extracts, 63–66
mechanical beetle, 154–55
Mechnikov, Ilya, 106, 107
medical education and students, 147
medical products, misinformation
 about, 142, 144
medications (drugs), 50–52, 100, 101

This book is also available as a Global Certified Accessible™ (GCA) ebook. ECW Press's ebooks are screen reader friendly and are built to meet the needs of those who are unable to read standard print due to blindness, low vision, dyslexia, or a physical disability.

Get the ebook free!*
*proof of purchase required

At ECW Press, we want you to enjoy our books in whatever format you like. If you've bought a print copy or an audiobook not purchased with a subscription credit, just send an email to ebook@ecwpress.com and include:

- the book title
- the name of the store where you purchased it
- a screenshot or picture of your order/receipt number and your name

A real person will respond to your email with your ePub attached. If you prefer to receive the ebook in PDF format, please let us know in your email.

Some restrictions apply. This offer is only valid for books already available in the ePub format. Some ECW Press books do not have an ePub format for us to send you. In those cases, we will let you know if a PDF format is available as an alternative. This offer is only valid for books purchased for personal use. At this time, this program is not offered on school or library copies.

Thank you for supporting an independently owned Canadian publisher with your purchase!